ELON MUSK

ALSO BY ASHLEE VANCE

Geek Silicon Valley:
The Inside Guide to Palo Alto, Stanford, Menlo Park, Mountain View,
Santa Clara, Sunnyvale, San Jose, San Francisco

ELON MUSK

HOW THE BILLIONAIRE CEO OF SPACEX AND TESLA IS SHAPING OUR FUTURE

ASHLEE VANCE

1 3 5 7 9 10 8 6 4 2

Virgin Books, an imprint of Ebury Publishing,
20 Vauxhall Bridge Road,
London SW1V 2SA

Virgin Books is part of the Penguin Random House group of companies
whose addresses can be found at global.penguinrandomhouse.com

Penguin
Random House
UK

First published in The United States by HarperCollins in 2015
First published in The United Kingdom by Virgin Books in 2015

www.eburypublishing.co.uk

A CIP catalogue record for this book is available from the British Library

Hardback ISBN: 9780753555620
Trade Paperback ISBN: 9780753555637

Printed and bound in India by Thomson Press India Ltd.

For Mum and Dad. Thanks for Everything.

CONTENTS

1

ELON'S WORLD

DO YOU THINK I'M INSANE?"

This question came from Elon Musk near the very end of a long dinner we shared at a high-end seafood restaurant in Silicon Valley. I'd gotten to the restaurant first and settled down with a gin and tonic, knowing Musk would—as ever—be late. After about fifteen minutes, Musk showed up wearing leather shoes, designer jeans, and a plaid dress shirt. Musk stands six foot one but ask anyone who knows him and they'll confirm that he seems much bigger than that. He's absurdly broad-shouldered, sturdy, and thick. You'd figure he would use this frame to his advantage and perform an alpha-male strut when entering a room. Instead, he tends to be almost sheepish. It's head tilted slightly down while walking, a quick handshake hello after reaching the table, and then butt in seat. From there, Musk needs a few minutes before he warms up and looks at ease.

Musk asked me to dinner for a negotiation of sorts. Eighteen months earlier, I'd informed him of my plans to write a book about him, and he'd informed me of his plans not to cooperate. His rejection stung but thrust me into dogged reporter mode. If I had to do this book without him, so be it. Plenty of people had left Musk's companies, Tesla Motors and SpaceX, and would talk, and I already knew a lot of his friends. The interviews followed one after another, month after month, and two hundred or so people into the process, I heard from Musk once again. He called me at home and declared that things could go one of two ways: he could make my life very difficult or he could help with the project after all. He'd be willing to cooperate if he could read the book before it went to publication, and could add footnotes throughout it. He would not meddle with my text, but he wanted the chance to set the record straight in spots that he deemed factually inaccurate. I understood where this was coming from. Musk wanted a measure of control over his life's story. He's also wired like a scientist and suffers mental anguish at the sight of a factual error. A mistake on a printed page would gnaw at his soul—forever. While I could understand his perspective, I could not let him read the book, for professional, personal, and practical reasons. Musk has his version of the truth, and it's not always the version of the truth that the rest of the world shares. He's prone to verbose answers to even the simplest of questions as well, and the thought of thirty-page footnotes seemed all too real. Still, we agreed to have dinner, chat all this out, and see where it left us.

Our conversation began with a discussion of public-relations people. Musk burns through PR staffers notoriously fast, and Tesla was in the process of hunting for a new communications chief. "Who is the best PR person in the world?" he asked in a very Muskian fashion. Then we talked about mutual acquaintances, Howard Hughes, and the Tesla factory. When the waiter stopped by to take our order, Musk asked for suggestions that would work with his low-carb diet. He settled on chunks of fried lobster soaked in black squid ink. The negotiation hadn't begun, and Musk was already dishing. He opened up about the major fear keeping him up at night: namely that Google's cofounder and CEO Larry Page might well have been building a fleet of artificial-intelligence-enhanced robots capable of destroying mankind. "I'm really worried about this," Musk said. It didn't make Musk feel any better that he and Page were very close friends and that he felt Page was fundamentally a well-intentioned person and not Dr. Evil. In fact, that was sort of the problem. Page's nice-guy nature left him assuming that the machines would forever do our bidding. "I'm not as optimistic," Musk said. "He could produce something evil by accident." As the food arrived, Musk consumed it. That is, he didn't eat it as much as he made it disappear rapidly with a few gargantuan bites. Desperate to keep Musk happy and chatting, I handed him a big chunk of steak from my plate. The plan worked . . . for all of ninety seconds. Meat. Hunk. Gone.

It took awhile to get Musk off the artificial intelligence doom-and-gloom talk and to the subject at hand. Then, as we drifted toward the book, Musk started to feel me out, probing exactly why it was that I wanted to write about him and calculating my intentions. When the moment presented itself, I moved in and seized the conversation. Some adrenaline released and mixed with the gin, and I launched into what was meant to be a forty-five-minute sermon about all the reasons Musk should let me burrow deep into his life and do so while getting exactly none of the controls he wanted in return. The speech revolved around the inherent limitations of footnotes, Musk coming off like a control freak and my journalistic integrity being compromised. To my great surprise, Musk cut me off after a couple of minutes and simply said, "Okay." One thing that Musk holds in the highest regard is resolve, and he respects people who continue on after being told no. Dozens of other journalists had asked him to help with a book

and transformed to suit humans. Musk fully intends to try and make this happen. Turning humans into space colonizers is his stated life's purpose. "I would like to die thinking that humanity has a bright future," he said. "If we can solve sustainable energy and be well on our way to becoming a multiplanetary species with a self-sustaining civilization on another planet—to cope with a worst-case scenario happening and extinguishing human consciousness—then," and here he paused for a moment, "I think that would be really good."

If some of the things that Musk says and does sound absurd, that's because on one level they very much are. On this occasion, for example, Musk's assistant had just handed him some cookies-and-cream ice cream with sprinkles on top, and he then talked earnestly about saving humanity while a blotch of the dessert hung from his lower lip.

Musk's ready willingness to tackle impossible things has turned him into a deity in Silicon Valley, where fellow CEOs like Page speak of him in reverential awe, and budding entrepreneurs strive "to be like Elon" just as they had been striving in years past to mimic Steve Jobs. Silicon Valley, though, operates within a warped version of reality, and outside the confines of its shared fantasy, Musk often comes off as a much more polarizing figure. He's the guy with the electric cars, solar panels, and rockets peddling false hope. Forget Steve Jobs. Musk is a sci-fi version of P. T. Barnum who has gotten extraordinarily rich by preying on people's fear and self-hatred. Buy a Tesla. Forget about the mess you've made of the planet for a while.

I'd long been a subscriber to this latter camp. Musk had struck me as a well-intentioned dreamer—a card-carrying member of Silicon Valley's techno-utopian club. This group tends to be a mix of Ayn Rand devotees and engineer absolutists who see their hyperlogical worldviews as the Answer for everyone. If we'd just get out of their way, they'd fix all our problems. One day, soon enough, we'll be able to download our brains to a computer, relax, and let their algorithms take care of everything. Much of their ambition proves inspiring and their works helpful. But the techno-utopians do get tiresome with their platitudes and their ability to prattle on for hours without saying much of substance. More disconcerting is their underlying message that humans are flawed and our humanity is an annoying burden that needs to be dealt with in due course. When I'd caught Musk at Silicon Valley events, his highfalutin talk often sounded straight out of the techno-utopian playbook. And, most annoyingly, his world-saving companies didn't even seem to be doing all that well.

Yet, in the early part of 2012, the cynics like me had to take notice of what Musk was actually accomplishing. His once-beleaguered companies were succeeding at unprecedented things. SpaceX flew a supply capsule to the International Space Station and brought it safely back to Earth. Tesla Motors delivered the Model S, a beautiful, all-electric sedan that took the automotive industry's breath away and slapped Detroit sober. These two feats elevated Musk to the rarest heights among business titans. Only Steve Jobs could claim similar achievements in two such different industries, sometimes putting out a new Apple product and a blockbuster Pixar movie in the same year. And yet, Musk was not done. He was also the chairman and largest shareholder of SolarCity, a booming solar energy company poised to file for an initial public offering. Musk had somehow delivered the biggest advances the space, automotive, and energy industries had seen in decades in what felt like one fell swoop.

It was in 2012 that I decided to see what Musk was like firsthand and to write a cover story about him for *Bloomberg Businessweek*. At this point in Musk's life, everything ran through his assistant/loyal appendage Mary Beth Brown. She invited me to visit what I've come to refer to as Musk Land.

Anyone arriving at Musk Land for the first time will have the same head-scratching experience. You're told to park at One Rocket Road in Hawthorne, where SpaceX has its HQ. It seems impossible that anything good could call Hawthorne home. It's a bleak part of Los Angeles County in which groupings of run-down houses, run-down shops, and run-down eateries surround huge, industrial complexes that appear to have been built during some kind of architectural Boring Rectangle movement. Did Elon Musk really stick his company in the middle of this dreck? Then, okay, things start to make more sense when you see one 550,000-square-foot rectangle painted an ostentatious hue of "Unity of Body, Soul, and Mind" white. This is the main SpaceX building.

It was only after going through the front doors of SpaceX that the grandeur of what this man had done became apparent. Musk had built an honest-to-God rocket factory in the middle of Los Angeles. And this factory was not making one rocket at a time. No. It was making many rockets—from scratch. The factory was a giant, shared work area. Near the back were massive delivery bays that allowed for the arrival of hunks of metal, which were transported to two-story-high welding machines. Over to one side were technicians in white coats making motherboards, radios, and other electronics. Other people were in a special, airtight glass chamber, building the capsules that rockets would take to the Space Station. Tattooed men in bandanas were blasting Van Halen and threading wires around rocket engines. There were completed bodies of rockets lined up one after the other ready to be placed on trucks. Still more rockets, in another part of the building, awaited coats of white paint. It was difficult to take in the entire factory at once. There were hundreds of bodies in constant motion whirring around a variety of bizarre machines.

This is just building number one of Musk Land. SpaceX had acquired several buildings that used to be part of a Boeing factory, which made the fuselages for 747s. One of these buildings has a curved roof and looks like an airplane hangar. It serves as the research, development, and design studio for Tesla. This is where the company came up with the look for the Model S sedan and its follow-on, the Model X SUV. In the parking lot outside the studio, Tesla has built one of its recharging stations where Los Angeles drivers can top up with electricity for free. The charging center is easy enough to spot because Musk has installed a white and red obelisk branded with the Tesla logo that sits in the middle of an infinity pool.

It was in my first interview with Musk, which took place at the design studio, that I began to get a sense of how he talked and operated. He's a confident guy, but does not always do a good job of displaying this. On initial encounter, Musk can come off as shy and borderline awkward. His South African accent remains present but fading, and the charm of it is not enough to offset the halting nature of Musk's speech pattern. Like many an engineer or physicist, Musk will pause while fishing around for exact phrasing, and he'll often go rumbling down an esoteric, scientific rabbit hole without providing any helping hands or simplified explanations along the way. Musk expects you to keep up. None of this is off-putting. Musk, in fact, will toss out plenty of jokes and can be downright charming. It's just that there's a sense of purpose and pressure hanging over any conversation with the man. Musk doesn't really shoot the shit. (It would end up taking about thirty hours of interviews for Musk to really loosen up and let me into a different, deeper level of his psyche and personality.)

Most high-profile CEOs have handlers all around them. Musk mostly moves about Musk Land on his own. This is not the guy who slinks into the restaurant. It's the guy who owns the joint and strides about with authority. Musk and I talked, as he made his way around the design studio's main floor, inspecting prototype parts and vehicles. At each station, employees rushed up to Musk and disgorged information. He listened intently, processed it, and nodded when satisfied. The people moved away and Musk moved to the next information dump. At one point, Tesla's design chief, Franz von Holzhausen, wanted Musk's take on some new tires and rims that had come in for the Model S and on the seating arrangements for the Model X. They spoke, and then they went into a back room where executives from a seller of high-end graphics software had prepared a presentation for Musk. They wanted to show off new 3-D rendering technology that would allow Tesla to tweak the finish of a virtual Model S and see in great detail how things like shadows and streetlights played off the car's body. Tesla's engineers really wanted the computing systems and needed Musk's sign-off. The men did their best to sell Musk on the idea while the sound of drills and giant industrial fans drowned out their shtick. Musk, wearing leather shoes, designer jeans, and a black T-shirt, which is essentially his work uniform, had to don 3-D goggles for the demonstration and seemed unmoved. He told them he'd think about it and then walked toward the source of the loudest noise—a workshop deep in the design studio where Tesla engineers were building the scaffolding for the thirty-foot decorative towers that go outside the charging stations. "That thing looks like it could survive a Category Five hurricane," Musk said. "Let's thin it up a bit." Musk and I eventually hop into his car—a black Model S—and zip back to the main SpaceX building. "I think there are probably too many smart people pursuing Internet stuff, finance, and law,"

announce it to the world in order for eager investors to fund your thought experiment. The whole goal was to make as much money as possible in the shortest amount of time because everyone knew on at least a subconscious level that reality had to set in eventually.

Valley denizens took very literally the cliché of working as hard as you play. People in their twenties, thirties, forties, and fifties were expected to pull all-nighters. Cubicles were turned into temporary homes, and personal hygiene was abandoned. Oddly enough, making Nothing appear to be Something took a lot of work. But when the time to decompress arrived, there were plenty of options for total debauchery. The hot companies and media powers of the time seemed locked in a struggle to outdo each other with ever-fancier parties. Old-line companies trying to look "with it" would regularly buy space at a concert venue and then order up some dancers, acrobats, open bars, and the Barenaked Ladies. Young technologists would show up to pound their free Jack and Cokes and snort their cocaine in porta-potties. Greed and self-interest were the only things that made any sense back then.

While the good times have been well chronicled, the subsequent bad times have been—unsurprisingly—ignored. It's more fun to reminiscence on irrational exuberance than the mess that gets left behind.

Let it be said for the record, then, that the implosion of the get-rich-quick Internet fantasy left San Francisco and Silicon Valley in a deep depression. The endless parties ended. The prostitutes no longer roamed the streets of the Tenderloin at 6 A.M. offering pre-commute love. ("Come on, honey. It's better than coffee!") Instead of the Barenaked Ladies, you got the occasional Neil Diamond tribute band at a trade show, some free T-shirts, and a lump of shame.

The technology industry had no idea what to do with itself. The dumb venture capitalists who had been taken during the bubble didn't want to look any dumber, so they stopped funding new ventures altogether. Entrepreneurs' big ideas were replaced by the smallest of notions. It was as if Silicon Valley had entered rehab en masse. It sounds melodramatic, but it's true. A populace of millions of clever people came to believe that they were inventing the future. Then . . . poof! Playing it safe suddenly became the fashionable thing to do.

The evidence of this malaise is in the companies and ideas formed during this period. Google had appeared and really started to thrive around 2002, but it was an outlier. Between Google and Apple's introduction of the iPhone in 2007, there's a wasteland of ho-hum companies. And the hot new things that were just starting out—Facebook and Twitter—certainly did not look like their predecessors—Hewlett-Packard, Intel, Sun Microsystems—that made physical products and employed tens of thousands of people in the process. In the years that followed, the goal went from taking huge risks to create new industries and grand new ideas, to chasing easier money by entertaining consumers and pumping out simple apps and advertisements. "The best minds of my generation are thinking about how to make people click ads," Jeff Hammerbacher, an early Facebook engineer, told me. "That sucks." Silicon Valley began to look an awful lot like Hollywood. Meanwhile, the consumers it served had turned inward, obsessed with their virtual lives.

One of the first people to suggest that this lull in innovation could signal a much larger problem was Jonathan Huebner, a physicist who works at the Pentagon's Naval Air Warfare Center in China Lake, California. Huebner is the *Leave It to Beaver* version of a merchant of death. Middle-aged, thin, and balding, he likes to wear a dirt-inspired ensemble of khaki pants, a brown-striped shirt, and a canvas khaki jacket. He has designed weapons systems since 1985, gaining direct insight into the latest and greatest technology around materials, energy, and software. Following the dot-com bust, he became miffed at the ho-hum nature of the supposed innovations crossing his desk. In 2005, Huebner delivered a paper, "A Possible Declining Trend in Worldwide Innovation," which was either an indictment of Silicon Valley or at least an ominous warning.

Huebner opted to use a tree metaphor to describe what he saw as the state of innovation. Man has already climbed past the trunk of the tree and gone out on its major limbs, mining most of the really big, game-changing ideas—the wheel, electricity, the airplane, the telephone, the transistor. Now we're left dangling near the end of the branches at the top of the tree and mostly just refining past inventions. To back up his point in the paper, Huebner showed that the frequency of life-changing inventions had started to slow. He also used data to prove that the number of patents filed per person had declined over time. "I think the probability of us discovering another top-one-hundred-type invention gets smaller and smaller," Huebner told me in an interview. "Innovation is a finite resource."

Huebner predicted that it would take people about five years to catch on to his thinking, and this forecast proved almost exactly right. Around 2010, Peter Thiel, the PayPal cofounder and early Facebook investor, began promoting the idea that the technology industry had let people down. "We wanted flying cars, instead we got 140 characters" became the tagline of his venture capital firm Founders Fund. In an essay called "What Happened to the Future," Thiel and his cohorts described how Twitter, its 140-character messages, and similar inventions have let the public down. He argued that science fiction, which once celebrated the future, has turned dystopian because people no longer have an optimistic view of technology's ability to change the world.

I'd subscribed to a lot of this type of thinking until that first visit to Musk Land. While Musk had been anything but shy about what he was up to, few people outside of his companies got to see the factories, the R&D centers, the machine shops, and to witness the scope of what he was doing firsthand. Here was a guy who had taken much of the Silicon Valley ethic behind moving quickly and running organizations free of bureaucratic hierarchies and applied it to improving big, fantastic machines and chasing things that had the potential to be the real breakthroughs we'd been missing.

By rights, Musk should have been part of the malaise. He jumped right into dot-com mania in 1995, when, fresh out of college, he founded a company called Zip2—a primitive Google Maps meets Yelp. That first venture ended up a big, quick hit. Compaq bought Zip2 in 1999 for $307 million. Musk made $22 million from the deal and poured almost all of it into his next venture, a start-up that would morph into PayPal. As the largest shareholder in PayPal, Musk became fantastically well-to-do when eBay acquired the company for $1.5 billion in 2002.

Instead of hanging around Silicon Valley and falling into the same funk as his peers, however, Musk decamped to Los Angeles. The conventional wisdom of the time said to take a deep breath and wait for the next big thing to arrive in due course. Musk rejected that logic by throwing $100 million into SpaceX, $70 million into Tesla, and $10 million into SolarCity. Short of building an actual money-crushing machine, Musk could not have picked a faster way to destroy his fortune. He became a one-man, ultra-risk-taking venture capital shop and doubled down on making super-complex physical goods in two of the most expensive places in the world, Los Angeles and Silicon Valley. Whenever possible, Musk's companies would make things from scratch and try to rethink much that the aerospace, automotive, and solar industries had accepted as convention.

With SpaceX, Musk is battling the giants of the U.S. military-industrial complex, including Lockheed Martin and Boeing. He's also battling nations—most notably Russia and China. SpaceX has made a name for itself as the low-cost supplier in the industry. But that, in and of itself, is not really good enough to win. The space business requires dealing with a mess of politics, back-scratching, and protectionism that undermines the fundamentals of capitalism. Steve Jobs faced similar forces when he went up against the recording industry to bring the iPod and iTunes to market. The crotchety Luddites in the music industry were a pleasure to deal with compared to Musk's foes who build weapons and countries for a living. SpaceX has been testing reusable rockets that can carry payloads to space and land back on Earth, on their launchpads, with precision. If the company can perfect this technology, it will deal a devastating blow to all of its competitors and almost assuredly push some mainstays of the rocket industry out of business while establishing the United States as the world leader for taking cargo and humans to space. It's a threat that Musk figures has earned him plenty of fierce enemies. "The list of people that would not mind if I was gone is growing," Musk said. "My family fears that the Russians will assassinate me."

With Tesla Motors, Musk has tried to revamp the way cars are manufactured and sold, while building out a worldwide fuel distribution network at the same time. Instead of hybrids, which in Musk lingo are suboptimal compromises, Tesla strives to make all-electric cars that people lust after and that push the limits of technology. Tesla does not sell these cars through dealers; it sells them on the Web and in Apple-like galleries located in high-end shopping centers. Tesla also does not anticipate making lots of money from servicing its vehicles, since electric cars do not require the oil changes and other maintenance procedures of traditional cars. The direct sales model embraced by Tesla stands as a major affront to car dealers used to haggling with their customers and making their profits from exorbitant maintenance fees. Tesla's recharging stations now run alongside many of the major highways in the United States, Europe, and Asia and can add hundreds of miles of oomph back to a car in about twenty minutes. These so-called supercharging stations are solar-powered, and Tesla owners pay nothing to refuel. While much of America's infrastructure decays, Musk is building a futuristic end-to-end transportation system that would allow the United States to leapfrog the rest of the world. Musk's vision, and, of late, execution seem to combine the best of Henry Ford and John D. Rockefeller.

With SolarCity, Musk has funded the largest installer and financer of solar panels for consumers and businesses. Musk helped come up with the idea for SolarCity and serves as its chairman, while his cousins Lyndon and Peter Rive run the company. SolarCity has managed to undercut dozens of utilities and become a large utility in its own right. During a time in which clean-tech businesses have gone bankrupt with alarming regularity, Musk has built two of the most successful clean-tech companies in the world. The Musk Co. empire of factories, tens of thousands of workers, and industrial might has incumbents on the run and has turned Musk into one of the richest men in the world, with a net worth around $10 billion.

The visit to Musk Land started to make a few things clear about how Musk had pulled all this off. While the "putting man on Mars" talk can strike some people as loopy, it gave Musk a unique rallying cry for his companies. It's the sweeping goal that forms a unifying principle over everything he does. Employees at all three companies are well aware of this and well aware that they're trying to achieve the impossible day in and day out. When Musk sets unrealistic goals, verbally abuses employees, and works them to the bone, it's understood to be—on some level—part of the Mars agenda. Some employees love him for this. Others loathe him but remain oddly loyal out of respect for his drive and mission. What Musk has developed that so many of the entrepreneurs in Silicon Valley lack is a meaningful worldview. He's the possessed genius on the grandest quest anyone has ever concocted. He's less a CEO chasing riches than a general marshaling troops to secure victory. Where Mark Zuckerberg wants to help you share baby photos, Musk wants to . . . well . . . save the human race from self-imposed or accidental annihilation.

The life that Musk has created to manage all of these endeavors is preposterous. A typical week starts at his mansion in Bel Air. On Monday, he works the entire day at SpaceX. On Tuesday, he begins at SpaceX, then hops onto his jet and flies to Silicon Valley. He spends a couple of days working at Tesla, which has its offices in Palo Alto and factory in Fremont. Musk does not own a home in Northern California and ends up staying at the luxe Rosewood hotel or at friends' houses. To arrange the stays with friends, Musk's assistant will send an e-mail asking, "Room for one?" and if the friend says, "Yes," Musk turns up at the door late at night. Most often he stays in a guest room, but he's also been known to crash on the couch after winding down with some video games. Then it's back to Los Angeles and SpaceX on Thursday. He shares custody of his five young boys—twins and triplets—with his ex-wife, Justine, and has them four days a week. Each year, Musk tabulates the amount of flight time he endures per week to help him get a sense of just how out of hand things are getting. Asked how he survives this schedule, Musk said, "I had a tough childhood, so maybe that was helpful."

During one visit to Musk Land, he had to squeeze our interview in before heading off for a camping trip at Crater Lake National Park in Oregon. It was almost 8 P.M. on a Friday, so Musk would soon be piling his boys and nannies into his private jet and then meeting drivers who would take him to his friends at the campsite; the friends would then help the Musk clan unpack and complete their pitch-black arrival. There would be a bit of hiking over the weekend. Then the relaxation would end. Musk would fly with the boys back to Los Angeles on Sunday afternoon. Then, he would take off on his own that evening for New York. Sleep. Hit the morning talk shows on Monday. Meetings. E-mail. Sleep. Fly back to Los Angeles Tuesday morning. Work at SpaceX. Fly to San Jose Tuesday afternoon to visit the Tesla Motors factory. Fly to Washington, D.C., that night and see President Obama. Fly back to Los Angeles Wednesday night. Spend a couple of days working at SpaceX. Then go to a weekend conference held by Google's chairman, Eric Schmidt, in Yellowstone. At this time, Musk had just split from his second wife, the actress Talulah Riley, and was trying to calculate if he could mix a personal life into all of this. "I think the time allocated to the businesses and the kids is going fine," Musk said. "I would like to allocate more time to dating, though. I need to find a girlfriend. That's why I need to carve out just a little more time. I think maybe even another five to ten—how much time does a woman want a week? Maybe ten hours? That's kind of the minimum? I don't know."

Musk rarely finds time to decompress, but when he does, the festivities are just as dramatic as the rest of his life. On his thirtieth birthday, Musk rented out a castle in England for about twenty people. From 2 A.M. until 6 A.M., they played a variation of hide-and-seek called sardines in which one person runs off and hides and everyone else looks for him. Another party occurred in Paris. Musk, his brother, and cousins found themselves awake at midnight and decided to bicycle through the city until 6 A.M. They slept all day and then boarded the Orient Express in the evening. Once again, they stayed up all night. The Lucent Dossier Experience—an avant-garde group of performers—were on the luxurious train, performing palm readings and acrobatics. When the train arrived in Venice the next day, Musk's family had dinner and then hung out on the patio of their hotel overlooking the Grand Canal until 9 A.M. Musk loves costume parties as well, and turned up at one dressed like a knight and using a parasol to duel a midget wearing a Darth Vader costume.

For one of his most recent birthdays, Musk invited fifty people to a castle—or at least the United States' best approximation of a castle—in Tarrytown, New York. This party had a Japanese steampunk theme, which is sort of like a sci-fi lover's wet dream—a mix of corsets, leather, and machine worship. Musk dressed as a samurai.

The festivities included a performance of *The Mikado*, a Victorian comic opera by Gilbert and Sullivan set in Japan, at a small theater in the heart of town. "I am not sure the Americans got it," said Riley, whom Musk remarried after his ten-hour-a-week dating plan failed. The Americans and everyone else did enjoy what followed. Back at the castle, Musk donned a blindfold, got pushed up against a wall, and held balloons in each hand and another between his legs. The knife thrower then went to work. "I'd seen him before, but did worry that maybe he could have an off day," Musk said. "Still, I thought, he would maybe hit one gonad but not both." The onlookers were stunned and frightened for Musk's safety. "That was bizarre," said Bill Lee, a technology investor and one of Musk's good friends. "But Elon believes in the science of things." One of the world's top sumo wrestlers showed up at the party along with some of his compatriots. A ring had been set up at the castle, and Musk faced off against the champion. "He was three hundred and fifty pounds, and they were not jiggly pounds," Musk said. "I went full adrenaline rush and managed to lift the guy off the ground. He let me win that first round and then beat me. I think my back is still screwed up."

Riley turned planning these types of parties for Musk into an art. She met Musk back in 2008, when his companies were collapsing. She watched him lose his entire fortune and get ridiculed by the press. She knows that the sting of these years remains and has combined with the other traumas in Musk's life—the tragic loss of an infant son and a brutal upbringing in South Africa—to create a tortured soul. Riley has gone to great lengths to make sure Musk's escapes from work and this past leave him feeling refreshed if not healed. "I try to think of fun things he has not done before where he can relax," Riley said. "We're trying to make up for his miserable childhood now."

Genuine as Riley's efforts might have been, they were not entirely effective. Not long after the Sumo party, I found Musk back at work at the Tesla headquarters in Palo Alto. It was a Saturday, and the parking lot was full of cars. Inside of the Tesla offices, hundreds of young men were at work—some of them designing car parts on computers and others conducting experiments with electronics equipment on their desks. Musk's uproarious laugh would erupt every few minutes and carry through the entire floor. When Musk came into the meeting room where I'd been waiting, I noted how impressive it was for so many people to turn up on a Saturday. Musk saw the situation in a different light, complaining that fewer and fewer people had been working weekends of late. "We've grown fucking soft," Musk replied. "I was just going to send out an e-mail. We're fucking soft." (A word of warning: There's going to be a lot of "fuck" in this book. Musk adores the word, and so do most of the people in his inner circle.)

This kind of declaration seems to fit with our impressions of other visionaries. It's not hard to imagine Howard Hughes or Steve Jobs chastising their workforce in a similar way. Building things—especially big things—is a messy business. In the two decades Musk has spent creating companies, he's left behind a trail of people who either adore or despise him. During the course of my reporting, these people lined up to give me their take on Musk and the gory details of how he and his businesses operate.

My dinners with Musk and periodic trips to Musk Land revealed a different set of possible truths about the man. He's set about building something that has the potential to be much grander than anything Hughes or Jobs produced. Musk has taken industries like aerospace and automotive that America seemed to have given up on and recast them as something new and fantastic. At the heart of this transformation are Musk's skills as a software maker and his ability to apply them to machines. He's merged atoms and bits in ways that few people thought possible, and the results have been spectacular. It's true enough that Musk has yet to have a consumer hit on the order of the iPhone or to touch more than one billion people like Facebook. For the moment, he's still making rich people's toys, and his budding empire could be an exploded rocket or massive Tesla recall away from collapse. On the other hand, Musk's companies have already accomplished far more than his loudest detractors thought possible, and the promise of what's to come has to leave hardened types feeling optimistic during their weaker moments. "To me, Elon is the shining example of how Silicon Valley might be able to reinvent itself and be more relevant than chasing these quick IPOs and focusing on getting incremental products out," said Edward Jung, a famed software engineer and inventor. "Those things are important, but they are not enough. We need to look at different models of how to do things that are longer term in nature and

2

AFRICA

THE PUBLIC FIRST MET ELON REEVE MUSK IN 1984. THE SOUTH AFRICAN trade publication *PC and Office Technology* published the source code to a video game Musk had designed. Called Blastar, the science-fiction-inspired space game required 167 lines of instructions to run. This was back in the day when early computer users were required to type out commands to make their machines do much of anything. In that context, Musk's game did not shine as a marvel of computer science but it certainly surpassed what most twelve-year-olds were kicking out at the time. Its coverage in the magazine netted Musk five hundred dollars and provided some early hints about his character. The Blastar spread on page 69 of the magazine shows that the young man wanted to go by the sci-fi-author-sounding name E. R. Musk and that he already had visions of grand conquests dancing in his head. The brief explainer states, "In this game you have to destroy an alien space freighter, which is carrying deadly Hydrogen Bombs and Status Beam Machines. This game makes good use of sprites and animation, and in this sense makes the listing worth reading." (As of this writing, not even the Internet knows what "status beam machines" are.)

A boy fantasizing about space and battles between good and evil is anything but amazing. A boy who takes these fantasies seriously is more remarkable. Such was the case with the young Elon Musk. By the middle of his teenage years, Musk had blended fantasy and reality to the point that they were hard to separate in his mind. Musk came to see man's fate in the universe as a personal obligation. If that meant pursuing cleaner energy technology or building spaceships to extend the human species's reach, then so be it. Musk would find a way to make these things happen. "Maybe I read too many comics as a kid," Musk said. "In the comics, it always seems like they are trying to save the world. It seemed like one should try to make the world a better place because the inverse makes no sense."

At around age fourteen, Musk had a full-on existential crisis. He tried to deal with it like many gifted adolescents do, turning to religious and philosophical texts. Musk sampled a handful of ideologies and then ended up more or less back where he had started, embracing the sci-fi lessons found in one of the most influential books in his life: *The Hitchhiker's Guide to the Galaxy*, by Douglas Adams. "He points out that one of the really tough things is figuring out what questions to ask," Musk said. "Once you figure out the question, then the answer is relatively easy. I came to the conclusion that really we should aspire to increase the scope and scale of human consciousness in order to better understand what questions to ask." The teenage Musk then arrived at his ultralogical mission statement. "The only thing that makes sense to do is strive for greater collective enlightenment," he said.

It's easy enough to spot some of the underpinnings of Musk's search for purpose. Born in 1971, he grew up in Pretoria, a large city in the northeastern part of South Africa, just an hour's drive from Johannesburg. The specter of apartheid was present throughout his childhood, as South Africa frequently boiled over with tension and violence. Blacks and whites clashed, as did blacks of different tribes. Musk turned four years old just days after the Soweto Uprising, in which hundreds of black students died while protesting decrees of the white government. For years South Africa faced sanctions imposed by other nations due to its racist policies. Musk had the luxury of traveling abroad during his childhood and would have gotten a flavor for how outsiders viewed South Africa.

But what had even more of an impact on Musk's personality was the white Afrikaner culture so prevalent in Pretoria and the surrounding areas. Hypermasculine behavior was celebrated and tough jocks were revered. While Musk enjoyed a level of privilege, he lived as an outsider whose reserved personality and geeky inclinations ran against the prevailing attitudes of the time. His notion that something about the world had gone awry received constant reinforcement, and Musk, almost from his earliest days, plotted an escape from his surroundings and dreamed of a place that would allow his personality and dreams to flourish. He saw America in its most clichéd form, as the land of opportunity and the most likely stage for making the realization of his dreams possible. This is how it came to pass that a lonesome, gawky South African boy who talked with the utmost sincerity about pursuing "collective enlightenment" ended up as America's most adventurous industrialist.

When Musk did finally reach the United States in his twenties, it marked a return to his ancestral roots. Family trees suggest that ancestors bearing the Swiss German surname of Haldeman on the maternal side of Musk's family left Europe for New York during the Revolutionary War. From New York, they spread out to the prairies of the Midwest—Illinois and Minnesota, in particular. "We had people that fought on both sides of the Civil War apparently and were a family of farmers," said Scott Haldeman, Musk's uncle and the unofficial family historian.

Throughout his childhood, boys teased Musk because of his unusual name. He earned the first part of it from his great-grandfather John Elon Haldeman, who was born in 1872[1] and grew up in Illinois before heading to Minnesota. There he would meet his wife, Almeda Jane Norman, who was five years younger. By 1902, the couple had settled down in a log cabin in the central Minnesota town of Pequot and given birth to their son Joshua Norman Haldeman, Musk's grandfather. He would grow up to become an eccentric and exceptional man and a model for Musk.*

Joshua Norman Haldeman is described as an athletic, self-reliant boy. In 1907, his family moved to the prairies of Saskatchewan, and his father died shortly thereafter when Joshua was just seven, leaving the boy to help run the house. He took to the wide-open land and picked up bronco horseback riding, boxing, and wrestling. Haldeman would break in horses for local farm-

---

* Two years after the birth of his son, John Elon began to show signs of diabetes. The condition amounted to a death sentence at the time and, despite being only thirty-two, John Elon learned that he would likely have six months or so to live. With a bit of nursing experience behind her, Almeda took it upon herself to discover an elixir or treatment that would extend John Elon's life. According to family lore, she hit on chiropractic procedures as an effective remedy, and John Elon lived for five years following the original diabetes diagnosis. The life-giving procedures established what would become an oddly rich chiropractic tradition in the Haldeman family. Almeda studied at a chiropractic school in Minneapolis and earned her doctor of chiropractic, or, D.C., degree in 1905. Musk's great-grandmother went on to set up her own clinic and, as far as anyone can tell, became the first chiropractor to practice in Canada.

ers, often hurting himself in the process, and he organized one of Canada's first rodeos. Family pictures show Joshua dressed in a decorative pair of chaps demonstrating his rope-spinning skills. As a teenager, Haldeman left home to get a degree from the Palmer School of Chiropractic in Iowa and then returned to Saskatchewan to become a farmer.

When the depression hit in the 1930s, Haldeman fell into a financial crisis. He could not afford to keep up with bank loans on his equipment and had five thousand acres of land seized. "From then on, Dad didn't believe in banks or holding on to money," said Scott Haldeman, who would go on to receive his chiropractic degree from the same school as his father and become one of the world's top spinal pain experts. After losing the farm around 1934, Haldeman lived something of a nomadic existence that his grandson would replicate in Canada decades later. Standing six feet, three inches, he did odd jobs as a construction worker and rodeo performer before settling down as a chiropractor.*

By 1948, Haldeman had married a Canadian dance studio instructor, Winnifred Josephine Fletcher, or Wyn, and built a booming chiropractic practice. That year, the family, which already included a son and a daughter, welcomed twin daughters Kaye and Maye, Musk's mother. The children lived in a three-story, twenty-room house that included a dance studio to let Wyn keep teaching students. Ever in search of something new to do, Haldeman had picked up flying and bought his own plane. The family gained some measure of notoriety as people heard about Haldeman and his wife packing their kids into the back of the single-engine craft and heading off on excursions all around

---

* Haldeman also entered politics, trying to start his own political party in Saskatchewan, publishing a newsletter, and espousing conservative, antisocialist ideas. He would later make an unsuccessful run for Parliament and chair the Social Credit Party.

North America. Haldeman would often show up at political and chiropractic meetings in the plane and later wrote a book with his wife called *The Flying Haldemans: Pity the Poor Private Pilot.*

Haldeman seemed to have everything going for him when, in 1950, he decided to give it all away. The doctor-cum-politician had long railed against government interference in the lives of individuals and had come to see the Canadian bureaucracy as too meddlesome. A man who forbade swearing, smoking, Coca-Cola, and refined flour at his house, Haldeman contended that the moral character of Canada had started to decline. Haldeman also possessed an enduring lust for adventure. And so, over the course of a few months, the family sold their house and dance and chiropractic practices and decided to move to South Africa—a place Haldeman had never been. Scott Haldeman remembers helping his father disassemble the family's Bellanca Cruisair (1948) airplane and put it into crates before shipping it to Africa. Once in South Africa, the family rebuilt the plane and used it to scour the country for a nice place to live, ultimately settling on Pretoria, where Haldeman set up a new chiropractic practice.

The family's spirit for adventure seemed to know no bounds. In 1952, Joshua and Wyn made a 22,000-mile round-trip journey in their plane, flying up through Africa to Scotland and Norway. Wyn served as the navigator and, though not a licensed pilot, would sometimes take over the flying duties. The couple topped this effort in 1954, flying 30,000 miles to Australia and back. Newspapers reported on the couple's trip, and they're believed to be the only private pilots to get from Africa to Australia in a single-engine plane.*

---

* The journey took them up the African coast, across the Arabian Peninsula, all the way through Iran, India, and Malaysia and then down the Timor Sea to Australia. It required one year of preparation just to secure all of the necessary visas and paperwork, and they suffered

When not up in the air, the Haldemans were out in the bush going on great, monthlong expeditions to find the Lost City of the Kalahari Desert, a supposed abandoned city in southern Africa. A family photo from one of these excursions shows the five children in the middle of the African bush. They have gathered around a large metal pot being warmed by the embers of a campfire. The children look relaxed as they sit in folding chairs, legs crossed and reading books. Behind them is the ruby-red Bellanca plane, a tent, and a car. The tranquility of the scene belies how dangerous these trips were. During one incident, the family's truck hit a tree stump and forced the bumper through the radiator. Stuck in the middle of nowhere with no means of communication, Joshua worked for three days to fix the truck, while the family hunted for food. At other times, hyenas and leopards would circle the campfire at night, and, one morning, the family woke to find a lion three feet away from their main table. Joshua grabbed the first object he could find—a lamp—waved it, and told the lion to go away. And it did.*

The Haldemans had a laissez-faire approach to raising their children, which would extend over the generations to Musk. Their kids were never punished, as Joshua believed they would intuit their way to proper behavior. When mom and dad went off on their tremendous flights, the kids were left at home. Scott Haldeman can't remember his father setting foot at his school a single time even though his son was captain of the rugby team

---

from constant stomach bugs and an erratic schedule along the way. "Dad passed out crossing the Timor Sea, and mum had to take over until they hit Australia. He woke up right before they landed," said Scott Haldeman. "It was fatigue."

* Both Joshua and Wyn were accomplished marksmen and won national shooting competitions. In the mid-1950s, they also tied for first place in the eight-thousand-mile Cape Town to Algiers Motor Rally, beating pros in their Ford station wagon.

and a prefect. "To him, that was all just anticipated," said Scott Haldeman. "We were left with the impression that we were capable of anything. You just have to make a decision and do it. In that sense, my father would be very proud of Elon."

Haldeman died in 1974 at the age of seventy-two. He'd been doing practice landings in his plane and didn't see a wire attached to a pair of poles. The wire caught the plane's wheels and flipped the craft, and Haldeman broke his neck. Elon was a toddler at the time. But throughout his childhood, Elon heard many stories about his grandfather's exploits and sat through countless slide shows that documented his travels and trips through the bush. "My grandmother told these tales of how they almost died several times along their journeys," Musk said. "They were flying in a plane with literally no instruments—not even a radio, and they had road maps instead of aerial maps, and some of those weren't even correct. My grandfather had this desire for adventure, exploration doing crazy things." Elon buys into the idea that his unusual tolerance for risk may well have been inherited directly from his grandfather. Many years after the last slide show, Elon tried to find and purchase the red Bellanca plane but could not locate it.

Maye Musk, Elon's mother, grew up idolizing her parents. In her youth, she was considered a nerd. She liked math and science and did well at the coursework. By the age of fifteen, however, people had taken notice of some of her other attributes. Maye was gorgeous. Tall with ash-blond hair, Maye had the high cheekbones and angular features that would make her stand out anywhere. A friend of the family ran a modeling school, and Maye took some courses. On the weekends, she did runway shows, magazine shoots, occasionally showed up at a senator's or ambassador's home for an event, and ended up as a finalist for Miss South Africa. (Maye has continued to model into her sixties, appearing on the covers of magazines like *New York* and *Elle* and in Beyoncé's music videos.)

Maye and Elon's father, Errol Musk, grew up in the same neighborhood. They met for the first time when Maye, born in 1948, was about eleven. Errol was the cool kid to Maye's nerd but had a crush on her for years. "He fell in love with me because of my legs and my teeth," said Maye. The two would date on and off throughout their time at university. And, according to Maye, Errol spent about seven years as a relentless suitor seeking her hand in marriage and eventually breaking her will. "He just never stopped proposing," she said.

Their marriage was complicated from the start. Maye became pregnant during the couple's honeymoon and gave birth to Elon on June 28, 1971, nine months and two days after her wedding day. While they may not have enjoyed marital bliss, the couple carved out a decent life for themselves in Pretoria. Errol worked as a mechanical and electrical engineer and handled large projects such as office buildings, retail complexes, residential subdivisions, and an air force base, while Maye set up a practice as a dietician. A bit more than a year after Elon's birth came his brother Kimbal, and soon thereafter came their sister Tosca.

Elon exhibited all the traits of a curious, energetic tot. He picked things up easily, and Maye, like many mothers do, pegged her son as brilliant and precocious. "He seemed to understand things quicker than the other kids," she said. The perplexing thing was that Elon seemed to drift off into a trance at times. People spoke to him, but nothing got through when he had a certain, distant look in his eyes. This happened so often that Elon's parents and doctors thought he might be deaf. "Sometimes, he just didn't hear you," said Maye. Doctors ran a series of tests on Elon, and elected to remove his adenoid glands, which can improve hearing in children. "Well, it didn't change," said

Maye. Elon's condition had far more to do with the wiring of his mind than how his auditory system functioned. "He goes into his brain, and then you just see he is in another world," Maye said. "He still does that. Now I just leave him be because I know he is designing a new rocket or something."

Other children did not respond well to these dreamlike states. You could do jumping jacks right beside Musk or yell at him, and he would not even notice. He kept right on thinking, and those around him judged that he was either rude or really weird. "I do think Elon was always a little different but in a nerdy way," Maye said. "It didn't endear him to his peers."

For Musk, these pensive moments were wonderful. At five and six, he had found a way to block out the world and dedicate all of his concentration to a single task. Part of this ability stemmed from the very visual way in which Musk's mind worked. He could see images in his mind's eye with a clarity and detail that we might associate today with an engineering drawing produced by computer software. "It seems as though the part of the brain that's usually reserved for visual processing—the part that is used to process images coming in from my eyes—gets taken over by internal thought processes," Musk said. "I can't do this as much now because there are so many things demanding my attention but, as a kid, it happened a lot. That large part of your brain that's used to handle incoming images gets used for internal thinking." Computers split their hardest jobs between two types of chips. There are graphics chips that deal with processing the images produced by a television show stream or video game and computational chips that handle general purpose tasks and mathematical operations. Over time, Musk has ended up thinking that his brain has the equivalent of a graphics chip. It allows him to see things out in the world, replicate them in his mind, and imagine how they might change or behave when interacting with other objects. "For images and numbers, I can process their interrelationships and algorithmic relationships," Musk said. "Acceleration, momentum, kinetic energy—how those sorts of things will be affected by objects comes through very vividly."

The most striking part of Elon's character as a young boy was his compulsion to read. From a very young age, he seemed to have a book in his hands at all times. "It was not unusual for him to read ten hours a day," said Kimbal. "If it was the weekend, he could go through two books in a day." The family went on numerous shopping excursions in which they realized mid-trip that Elon had gone missing. Maye or Kimbal would pop into the nearest bookstore and find Elon somewhere near the back sitting on the floor and reading in one of his trancelike states.

As Elon got older, he would take himself to the bookstore when school ended at 2 P.M. and stay there until about 6 P.M., when his parents returned home from work. He plowed through fiction books and then comics and then nonfiction titles. "Sometimes they kicked me out of the store, but usually not," Elon said. He listed *The Lord of the Rings*, Isaac Asimov's Foundation series, and Robert Heinlein's *The Moon Is a Harsh Mistress* as some of his favorites, alongside *The Hitchhiker's Guide to the Galaxy*. "At one point, I ran out of books to read at the school library and the neighborhood library," Musk said. "This is maybe the third or fourth grade. I tried to convince the librarian to order books for me. So then, I started to read the *Encyclopaedia Britannica*. That was so helpful. You don't know what you don't know. You realize there are all these things out there."

Elon, in fact, churned through two sets of encyclopedias—a feat that did little to help him make friends. The boy had a photographic memory, and the encyclopedias turned him into a fact factory. He came off as a classic know-it-all. At the dinner table, Tosca would wonder aloud about the distance from Earth to the

Moon. Elon would spit out the exact measurement at perigee and apogee. "If we had a question, Tosca would always say, 'Just ask genius boy,'" Maye said. "We could ask him about anything. He just remembered it." Elon cemented his bookworm reputation through his clumsy ways. "He's not very sporty," said Maye.

Maye tells the story of Elon playing outside one night with his siblings and cousins. When one of them complained of being frightened by the dark, Elon pointed out that "dark is merely the absence of light," which did little to reassure the scared child. As a youngster, Elon's constant yearning to correct people and his abrasive manner put off other kids and added to his feelings of isolation. Elon genuinely thought that people would be happy to hear about the flaws in their thinking. "Kids don't like answers like that," said Maye. "They would say, 'Elon, we are not playing with you anymore.' I felt very sad as a mother because I think he wanted friends. Kimbal and Tosca would bring home friends, and Elon wouldn't, and he would want to play with them. But he was awkward, you know." Maye urged Kimbal and Tosca to include Elon. They responded as kids will. "But Mom, he's not fun." As he got older, however, Elon would have strong, affection-ate attachments to his siblings and cousins—his mother's sister's sons. Though he kept to himself at school, Elon had an outgoing nature with members of his family and eventually took on the role of elder and chief instigator among them.

For a while, life inside the Musk household was quite good. The family owned one of the biggest houses in Pretoria thanks to the success of Errol's engineering business. There's a portrait of the three Musk children taken when Elon was about eight years old that shows three blond, fit children sitting next to each other on a brick porch with Pretoria's famous purple jacaranda trees in the background. Elon has large, rounded cheeks and a broad smile.

Then, not long after the photo was taken, the family fell apart. His parents separated and divorced within the year. Maye moved with the kids to the family's holiday home in Durban, on South Africa's eastern coast. After a couple of years of this arrangement, Elon decided he wanted to live with his father. "My father seemed sort of sad and lonely, and my mom had three kids, and he didn't have any," Musk said. "It seemed unfair." Some members of Musk's family have bought into this idea that Elon's logical nature propelled him, while others claim that his father's mother, Cora, exerted a lot of pressure on the boy. "I could not understand why he would leave this happy home I made for him—this really happy home," said Maye. "But Elon is his own person." Justine Musk, Elon's ex-wife and the mother of his five boys, theorized that Elon identified more with the alpha male of the house and wasn't bothered by the emotional aspect of the decision. "I don't think he was particularly close with either parent," Justine said, while describing the Musk clan overall as being cool and the opposite of doting. Kimbal later opted to live with Errol as well, saying simply that by nature a son wants to live with his father.

Whenever the topic of Errol arrives, members of Elon's family clam up. They're in agreement that he was not a pleasant man to be around but have declined to elaborate. Errol has since been remarried, and Elon has two, younger half sisters of whom he's quite protective. Elon and his siblings seem determined not to bad-mouth Errol publicly, so as not to upset the sisters.

The basics are as follows: Errol's side of the family has deep South African roots. The Musk clan can trace its presence in the country back about two hundred years and claim an entry in Pretoria's first phone book. Errol's father, Walter Henry James Musk, was an army sergeant. "I remember him almost never talking," Elon said. "He would just drink whiskey and be grumpy and was very good at doing crossword puzzles." Cora Amelia Musk, Errol's mother, was born in England to a family famed for its intellectual genes. She embraced both the spotlight and her grandchildren. "Our grandmother had this very dominant personality and was quite an enterprising woman," said Kimbal. "She was a very big influence in our lives." Elon considered his relationship with Cora—or Nana, as he called her—particularly tight. "After the divorce, she took care of me quite a lot," he said. "She would pick me up from school, and I would hang out with her playing Scrabble and that type of thing."

On the surface, life at Errol's house seemed grand. He had plenty of books for Elon to read from cover to cover and money to buy a computer and other objects that Elon desired. Errol took his children on numerous trips overseas. "It was an amazingly fun time," said Kimbal. "I have a lot of fun memories from that." Errol also impressed the kids with his intellect and dealt out some practical lessons. "He was a talented engineer," Elon said. "He knew how every physical object worked." Both Elon and Kimbal were required to go to the sites of Errol's engineering jobs and learn how to lay bricks, install plumbing, fit windows, and put in electrical wiring. "There were fun moments," Elon said.

Errol was what Kimbal described as "ultra-present and very intense." He would sit Elon and Kimbal down and lecture at them for three to four hours without the boys being able to respond. He seemed to delight in being hard on the boys and sucked the fun out of common childhood diversions. From time to time, Elon tried to convince his dad to move to America and often talked about his intentions to live in the United States later in life. Errol countered such dreams by trying to teach Elon a lesson. He sent the housekeepers away and had Elon do all the chores to let him know what it was like "to play American."

While Elon and Kimbal declined to provide an exact recounting, they clearly experienced something awful and profound during those years with their father. They both talk about having had to endure a difficult childhood with some bad moments. "He definitely has serious chemical stuff," said Kimbal. "Which I am sure Elon and I have inherited. It was a very emotionally challenging upbringing, but it made us who we are today." Maye bristled when the subject of Errol came up, saying that she did not get on with him. "I don't want to tell stories. You know, you just don't talk about it. There are kids and grandkids involved."

When asked to chat about Elon, Errol responded via e-mail: "Elon was a very independent and focused child at home with me. He loved computer science before anyone even knew what it was in South Africa and his ability was widely recognized by the time he was 12 years old. Elon and his brother Kimbal's activities as children and young men were so many and varied that it's difficult to name just one, as they travelled together with me extensively in S. Africa and the world at large, visiting all the continents regularly from the age of six onwards. Elon and his brother and sister were and continue to be exemplary, in every way a father could want. I'm very proud of what Elon's accomplished."

Errol copied Elon on this e-mail, and Elon warned me off corresponding with his father, insisting that his father's take on past events could not be trusted. "He is an odd duck," Musk said. But, when pressed for more information, Musk dodged. "It would certainly be accurate to say that I did not have a good childhood," he said. "It may sound good. It was not absent of good, but it was not a happy childhood. It was like misery. He's good at making life miserable—that's for sure. He can take any situation no matter how good it is and make it bad. He's not a happy man. I don't know . . . fuck . . . I don't know how someone becomes like he is. It would just cause too much trouble to tell you any more." Elon and Justine have vowed that their children will not be allowed to meet Errol.

When Elon was nearly ten years old, he saw a computer for the first time, at the Sandton City Mall in Johannesburg. "There was an electronics store that mostly did hi-fi-type stuff, but then, in one corner, they started stocking a few computers," Musk said. He felt awed right away—"It was like, 'Whoa. Holy shit!' "—by this machine that could be programmed to do a person's bidding. "I had to have that and then hounded my father to get the computer," Musk said. Soon he owned a Commodore VIC-20, a popular home machine that went on sale in 1980. Elon's computer arrived with five kilobytes of memory and a workbook on the BASIC programming language. "It was supposed to take like six months to get through all the lessons," Elon said. "I just got super OCD on it and stayed up for three days with no sleep and did the entire thing. It seemed like the most super-compelling thing I had ever seen." Despite being an engineer, Musk's father was something of a Luddite and dismissive of the machine. Elon recounted that "he said it was just for games and that you'd never be able to do real engineering on it. I just said, 'Whatever.'"

While bookish and into his new computer, Elon quite often led Kimbal and his cousins (Kaye's children) Russ, Lyndon, and Peter Rive on adventures. They dabbled one year in selling Easter eggs door-to-door in the neighborhood. The eggs were not well decorated, but the boys still marked them up a few hundred percent for their wealthy neighbors. Elon also spearheaded their work with homemade explosives and rockets. South Africa did not have the Estes rocket kits popular among hobbyists, so Elon would create his own chemical compounds and put them inside of canisters. "It is remarkable how many things you can get to explode," Elon said. "Saltpeter, sulfur, and charcoal are the basic ingredients for gunpowder, and then if you combine a strong acid with a strong alkaline, that will generally release a lot of energy. Granulated chlorine with brake fluid—that's quite impressive. I'm lucky I have all my fingers." When not handling explosives, the boys put on layers of clothing and goggles and shot each other with pellet guns. Elon and Kimbal raced dirt bikes against each other in sandlots until Kimbal flew off his bike one day and hurtled into a barbed wire fence.

As the years went on, the cousins took their entrepreneurial pursuits more seriously, even attempting at one point to start a video arcade. Without any parents knowing, the boys picked out a spot for their arcade, got a lease, and started navigating the permit process for their business. Eventually, they had to get someone over eighteen to sign a legal document, and neither the Rives' father nor Errol would oblige. It would take a couple of decades, but Elon and the Rives would eventually go into business together.

The boys' most audacious exploits may have been their trips between Pretoria and Johannesburg. During the 1980s, South Africa could be a terribly violent place, and the thirty-five-mile train trip linking Pretoria and Johannesburg stood out as one of the world's more dangerous rides. Kimbal counted the train journeys as formative experiences for him and Elon. "South Africa was not a happy-go-lucky place, and that has an impact on you. We saw some really rough stuff. It was part of an atypical upbringing—just this insane set of experiences that changes how you view risk. You don't grow up thinking getting a job is the hard part. That's not interesting enough."

The boys ranged in age from about thirteen to sixteen and chased a mix of parties and geeky exploits in Johannesburg. During one jaunt, they went to a Dungeons & Dragons tournament. "That was us being nerd master supremes," Musk said. All of the boys were into the role-playing game, which requires someone to help set the mood for a contest by imagining and then describing a scene. "You have entered a room, and there is a chest in the corner. What will you do? . . . You open the chest. You've sprung a trap. Dozens of goblins are on the loose." Elon excelled at this Dungeon Master role and had memorized the texts detailing the powers of monsters and other characters. "Under Elon's leadership, we played the role so well and won the tournament," said Peter Rive. "Winning requires this incredible imagination, and Elon really set the tone for keeping people captivated and inspired."

The Elon that his peers encountered at school was far less inspirational. Throughout middle and high school, Elon bounced around a couple of institutions. He spent the equivalent of eighth and ninth grades at Bryanston High School. One afternoon Elon and Kimbal were sitting at the top of a flight of concrete stairs eating when a boy decided to go after Elon. "I was basically hiding from this gang that was fucking hunting me down for God knows fucking why. I think I accidentally bumped this guy at assembly that morning and he'd taken some huge offense at that." The boy crept up behind Musk, kicked him in the head, and then shoved him down the stairs. Musk tumbled down the entire flight, and a handful of boys pounced on him, some of them kicking Musk in the side and the ringleader bashing his head against the ground. "They were a bunch of fucking psychos," Musk said. "I blacked out." Kimbal watched in horror and feared for Elon's life. He rushed down the stairs to find Elon's face bloodied and swollen. "He looked like someone who had just been in the boxing ring," Kimbal said. Elon then went to the hospital. "It was about a week before I could get back to school," Musk said. (During a news conference in 2013, Elon disclosed that he'd had a nose job to deal with the lingering effects of this beating.)

For three or four years, Musk endured relentless hounding at the hands of these bullies. They went so far as to beat up a boy that Musk considered his best friend until the child agreed to stop hanging out with Musk. "Moreover, they got him—they got my best fucking friend—to lure me out of hiding so they could beat me up," Musk said. "And that fucking hurt." While telling this part of the story, Musk's eyes welled up and his voice quivered. "For some reason, they decided that I was it, and they were going to go after me nonstop. That's what made growing up difficult. For a number of years, there was no respite. You get chased around by gangs at school who tried to beat the shit out of me, and then I'd come home, and it would just be awful there as well. It was just like nonstop horrible."

Musk spent the latter stages of his high school career at Pretoria Boys High School, where a growth spurt and the generally better behavior of the students made life more bearable. While a public school by definition, Pretoria Boys has functioned more like a private school for the last hundred years. It's the place you send a young man to get him ready to attend Oxford or Cambridge.

The boys from Musk's class remember him as a likable, quiet, unspectacular student. "There were four or five boys that were considered the very brightest," said Deon Prinsloo, who sat behind Elon in some classes. "Elon was not one of them." Such comments were echoed by a half dozen boys who also noted that Musk's lack of interest in sports left him isolated in the midst of an athletics-obsessed culture. "Honestly, there were just no signs that he was going to be a billionaire," said Gideon Fourie, another classmate. "He was never in a leadership position at school. I was rather surprised to see what has happened to him."

While Musk didn't have any close friends at school, his eccentric interests did leave an impression. One boy—Ted Wood—remembered Musk bringing model rockets to school and blasting them off during breaks. This was not the only hint of his aspirations. During a science-class debate, Elon gained attention for railing against fossil fuels in favor of solar power—an almost sacrilegious stance in a country devoted to mining the earth's natural resources. "He always had firm views on things," said Wood. Terency Beney, a classmate who stayed in touch with Elon over the years, claimed that Musk had started fantasizing about colonizing other planets in high school as well.

In another nod to the future, Elon and Kimbal were chatting during a class break outdoors when Wood interrupted them and asked what they were going on about. "They said, 'We are talking about whether there is a need for branch banking in the financial industry and whether we will move to paperless banking.' I remember thinking that was such an absurd comment to make. I said, 'Yeah, that's great.'"*

While Musk might not have been among the academic elite in his class, he was among a handful of students with the grades and self-professed interest to be selected for an experimental computer program. Students were plucked out of a number of schools and brought together to learn the BASIC, COBOL, and Pascal programming languages. Musk continued to augment these technological leanings with his love of science fiction and fantasy and tried his hand at writing stories that involved dragons and supernatural beings. "I wanted to write something like *Lord of the Rings*," he said.

Maye viewed these high school years through a mother's eyes and recounted plenty of tales of Musk performing spectacular academic feats. The video game he wrote, she said, impressed

---

* Musk couldn't remember this particular conversation. "I think they might be having creative recollection," he said. "It's possible. I had lots of esoteric conversations the last couple years of high school, but I was more concerned about general technology than banking."

much older, more experienced techies. He aced math exams well beyond his years. And he had that incredible memory. The only reason he did not outrank the other boys was a lack of interest in the work prescribed by the school.

As Musk saw it, "I just look at it as 'What grades do I need to get where I want to go?' There were compulsory subjects like Afrikaans, and I just didn't see the point of learning that. It seemed ridiculous. I'd get a passing grade and that was fine. Things like physics and computers—I got the highest grade you can get in those. There needs to be a reason for a grade. I'd rather play video games, write software, and read books than try and get an A if there's no point in getting an A. I can remember failing subjects in like fourth and fifth grade. Then, my mother's boyfriend told me I'd be held back if I didn't pass. I didn't actually know you had to pass the subjects to move to the next grade. I got the best grades in class after that."

At seventeen, Musk left South Africa for Canada. He has recounted this journey quite often in the press and typically leans on two descriptions of the motivation for his flight. The short version is that Musk wanted to get to the United States as quickly as possible and could use Canada as a pit stop via his Canadian ancestry. The second go-to story that Musk relies on has more of a social conscience. South Africa required military service at the time. Musk wanted to avoid joining the military, he has said, because it would have forced him to participate in the apartheid regime.

What rarely gets mentioned is that Musk attended the University of Pretoria for five months before heading off on his grand adventure. He began pursuing physics and engineering but put lackluster effort into the work and soon dropped out of school. Musk characterized the time at university as just something to do while he awaited his Canadian documentation. In addition to being an inconsequential part of his life, Musk lazing through school to avoid South Africa's required military service rather undermines the tale of a brooding, adventurous youth that he likes to tell, which is likely why the stint at the University of Pretoria never seems to come up.

There's no question, though, that Musk had been pining to get to the United States on a visceral level for a long time. Musk's early inclination toward computers and technology had fostered an intense interest in Silicon Valley, and his trips overseas had reinforced the idea that America was the place to get things done. South Africa, by contrast, presented far less opportunity for an entrepreneurial soul. As Kimbal put it, "South Africa was like a prison for someone like Elon."

Musk's opportunity to flee arrived with a change in the law that allowed Maye to pass her Canadian citizenship to her children. Musk immediately began researching how to complete the paperwork for this process. It took about a year to receive the approvals from the Canadian government and to secure a Canadian passport. "That's when Elon said, 'I'm leaving for Canada,'" Maye said. In these pre-Internet days, Musk had to wait three agonizing weeks to get a plane ticket. Once it arrived, and without flinching, he left home for good.

3

CANADA

MUSK'S GREAT ESCAPE TO CANADA WAS NOT WELL THOUGHT OUT. HE knew of a great-uncle in Montreal, hopped on a flight and hoped for the best. Upon landing in June 1988, Musk found a pay phone and tried to use directory assistance to find his uncle. When that didn't work, he called his mother collect. She had bad news. Maye had sent a letter to the uncle before Musk left and received a reply while her son was in transit. The uncle had gone to Minnesota, meaning Musk had nowhere to stay. Bags in hand, Musk headed for a youth hostel.

After spending a few days in Montreal exploring the city, Musk tried to come up with a long-term plan. Maye had family scattered all across Canada, and Musk began reaching out to them. He bought a countrywide bus ticket that let him hop on and off as he pleased for one hundred dollars, and opted to head to Saskatchewan, the former home of his grandfather. After a 1,900-mile bus ride, he ended up in Swift Current, a town of fifteen thousand people. Musk called a second cousin out of the blue from the bus station and hitched a ride to his house.

Musk spent the next year working a series of odd jobs around Canada. He tended vegetables and shoveled out grain bins at a cousin's farm located in the tiny town of Waldeck. Musk celebrated his eighteenth birthday there, sharing a cake with the family he'd just met and a few strangers from the neighborhood. After that, he learned to cut logs with a chain saw in Vancouver, British Columbia. The hardest job Musk took came after a visit to the unemployment office. He inquired about the job with the best wage, which turned out to be a gig cleaning the boiler room of a lumber mill for eighteen dollars an hour. "You have to put on this hazmat suit and then shimmy through this little tunnel that you can barely fit in," Musk said. "Then, you have a shovel and you take the sand and goop and other residue, which is still steaming hot, and you have to shovel it through the same hole you came through. There is no escape. Someone else on the other side has to shovel it into a wheelbarrow. If you stay in there for more than thirty minutes, you get too hot and die." Thirty people started out at the beginning of the week. By the third day, five people were left. At the end of the week, it was just Musk and two other men doing the work.

As Musk made his way around Canada, his brother, sister, and mother were figuring out how to get there as well.* When Kimbal and Elon eventually reunited in Canada, their headstrong, playful natures bloomed. Elon ended up enrolling at Queen's Univer-

---

* When Maye went to Canada to check out places to live, a fourteen-year-old Tosca seized the moment and put the family house in South Africa up for sale. "She had sold my car as well and was in the midst of putting our furniture up for sale, too," Maye said. "When I got back, I asked her why. She said, 'There is no need to delay. We are getting out of here.'"

sity in Kingston, Ontario, in 1989. (He picked Queen's over the University of Waterloo because he felt there were more good-looking women at Queen's.)[2] Outside of his studies, Elon would read the newspaper alongside Kimbal, and the two of them would identify interesting people they would like to meet. They then took turns cold-calling these people to ask if they were available to have lunch. Among the harassed was the head of marketing for the Toronto Blue Jays baseball team, a business writer for the *Globe and Mail*, and a top executive at the Bank of Nova Scotia, Peter Nicholson. Nicholson remembered the boys' call well. "I was not in the habit of getting out-of-the-blue requests," he said. "I was perfectly prepared to have lunch with a couple of kids that had that kind of gumption." It took six months to get on Nicholson's calendar, but, sure enough, the Musk brothers made a three-hour train ride and showed up on time.

Nicholson's first exposure to the Musk brothers left him with an impression many would share. Both presented themselves well and were polite. Elon, though, clearly came off as the geekier, more awkward counterpoint to the charismatic, personable Kimbal. "I became more impressed and fascinated as I talked to them," Nicholson said. "They were so determined." Nicholson ended up offering Elon a summer internship at the bank and became his trusted advisor.

Not long after their initial meeting, Elon invited Peter Nicholson's daughter Christie to his birthday party. Christie showed up at Maye's Toronto apartment with a jar of homemade lemon curd in hand and was greeted by Elon and about fifteen other people. Elon had never met Christie before, but he went right up to her and led her to a couch. "Then, I believe the second sentence out of his mouth was 'I think a lot about electric cars,'" Christie said. "And then he turned to me and said, 'Do you think about electric cars?'" The conversation left Christie, who is now a science writer, with the distinct impression that Musk was handsome, affable, and a tremendous nerd. "For whatever reason, I was so struck by that moment on the sofa," she said. "You could tell that this person was very different. He captivated me in that way."

With her angular features and blond hair, Christie fit Musk's type, and the two stayed in touch during Musk's time in Canada. They never really dated, but Christie found Musk interesting enough to have lengthy conversations with him on the phone. "One night he told me, 'If there was a way that I could not eat, so I could work more, I would not eat. I wish there was a way to get nutrients without sitting down for a meal.' The enormity of his work ethic at that age and his intensity jumped out. It seemed like one of the more unusual things I had ever heard."

A deeper relationship during this stint in Canada arose between Musk and Justine Wilson, a fellow student at Queen's. Leggy with long, brown hair, Wilson radiated romance and sexual energy. Justine had already fallen in love with an older man and then ditched him to go to college. Her next conquest was meant to wear a leather jacket and be a damaged, James Dean sort. As fortune would have it, however, the clean-cut, posh-sounding Musk spotted Wilson on campus and went right to work trying to date her. "She looked pretty great," Musk said. "She was also smart and this intellectual with sort of an edge. She had a black belt in tae kwon do and was semi-bohemian and, you know, like the hot chick on campus." He made his first move just outside of her dorm, where he pretended to have bumped into her by accident and then reminded her that they had met previously at a party. Justine, only one week into school, agreed to Musk's proposal of an ice cream date. When he arrived to pick up Wilson, Musk found a note on the dorm room door, notifying him that he'd been stood up. "It said that she had to go study for an exam and couldn't make it and that she was sorry," Musk said. Musk then hunted down Justine's best friend and did some research, asking where Justine usually studied and what her favorite flavor of ice cream was. Later, as Justine hid in the student center studying Spanish, Musk appeared behind her with a couple of melting chocolate chip ice cream cones in hand.

Wilson had dreamed of having a torrid romance with a writer. "I wanted to be Sylvia and Ted," she said. What she fell for instead was a relentless, ambitious geek. The pair attended the same abnormal-psychology class and compared their grades following an exam. Justine notched a 97, Musk a 98. "He went back to the professor, and talked his way into the two points he lost and got a hundred," Justine said. "It felt like we were always competing." Musk had a romantic side as well. One time he sent Wilson a dozen roses, each with its own note, and he also gifted Wilson a copy of *The Prophet* filled with handwritten romantic musings. "He can sweep you off your feet," Justine said.

During their university years, the two youngsters were off and on, with Musk having to work hard to keep the relationship going. "She was hip and dated the coolest guys and wasn't interested in Elon at all," Maye said. "So that was hard on him." Musk pursued a couple of other girls, but kept returning to Justine. Any time she acted cool toward him, Musk responded with his usual show of force. "He would call very insistently," she said. "You always knew it was Elon because the phone would never stop ringing. The man does not take no for an answer. You can't blow him off. I do think of him as the Terminator. He locks his gaze on to something and says, 'It shall be mine.' Bit by bit, he won me over."

College suited Musk. He worked on being less of a know-it-all, while also finding a group of people who respected his intellectual abilities. The university students were less inclined to laugh off or deride his opinionated takes on energy, space, and whatever else was captivating him at the moment. Musk had found people who responded to his ambition rather than mocking it, and he fed on this environment.

Navaid Farooq, a Canadian who grew up in Geneva, ended up in Musk's freshman-year dormitory in the fall of 1990. Both men were placed in the international section where a Canadian student would get paired with a student from overseas. Musk sort of broke the system, since he technically counted as a Canadian but knew almost nothing about his surroundings. "I had a roommate from Hong Kong, and he was a really nice guy," Musk said. "He religiously attended every lecture, which was helpful, since I went to the least number of classes possible." For a time, Musk sold computer parts and full PCs in the dorm to make some extra cash. "I could build something to suit their needs like a tricked-out gaming machine or a simple word processor that cost less than what they could get in a store," Musk said. "Or if their computer didn't boot properly or had a virus, I'd fix it. I could pretty much solve any problem." Farooq and Musk bonded over their backgrounds living abroad and a shared interest in strategy board games. "I don't think he makes friends easily, but he is very loyal to those he has," Farooq said. When the video game Civilization was released, the college chums spent hours building their empire, much to the dismay of Farooq's girlfriend, who was forgotten in another room. "Elon could lose himself for hours on end," Farooq said. The students also relished their loner lifestyles. "We are the kinds of people that can be by ourselves at a party and not feel awkward," Farooq said. "We can think to ourselves and not feel socially weird about it."

Musk was more ambitious in college than he'd been in high school. He studied business, competed in public speaking contests, and began to display the brand of intensity and competitiveness that marks his behavior today. After one economics exam, Musk, Farooq, and some other students in class came back to the dorms and began comparing notes to try to ascertain how well they did on the test. It soon became clear that Musk had a firmer grasp on the material than anyone else. "This was a group of fairly high achievers, and Elon stood way outside of the bell curve," Farooq said. Musk's intensity has continued to be a constant in their long relationship. "When Elon gets into something, he develops just this different level of interest in it than other people. That is what differentiates Elon from the rest of humanity."

In 1992, having spent two years at Queen's, Musk transferred to the University of Pennsylvania on a scholarship. Musk saw the Ivy League school as possibly opening some additional doors and went off in pursuit of dual degrees—first an economics degree from the Wharton School and then a bachelor's degree in physics. Justine stayed at Queen's, pursuing her dream of becoming a writer, and maintained a long-distance relationship with Musk. Now and again, she would visit him, and the two would sometimes head off to New York for a romantic weekend.

Musk blossomed even more at Penn, and really started to feel comfortable while hanging out with his fellow physics students. "At Penn, he met people that thought like him," Maye said. "There were some nerds there. He so enjoyed them. I remember going for lunch with them, and they were talking physics things. They were saying, 'A plus B equals pi squared' or whatever. They would laugh out loud. It was cool to see him so happy." Once again, however, Musk did not make many friends among the broader school body. It's difficult to find former students who remember him being there at all. But he did make one very close friend named Adeo Ressi, who would go on to be a Silicon Valley entrepreneur in his own right and is to this day as tight with Elon as anyone.

Ressi is a lanky guy well over six feet tall and possesses an eccentric air. He was the artistic, colorful foil to the studious, more buttoned-up Musk. Both of the young men were transfer students and ended up being placed in the funky freshman dorm. The lackluster social scene did not live up to Ressi's expectations, and he talked Musk into renting a large house off campus. They got the ten-bedroom home relatively cheap, since it was a frat house that had gone unrented. During the week, Musk and Ressi would study, but as the weekend approached, Ressi, in particular, would transform the house into a nightclub. He covered the windows with trash bags to make it pitch black inside and decorated the walls with bright paints and whatever objects he could find. "It was a full-out, unlicensed speakeasy," Ressi said. "We would have as many as five hundred people. We would charge five dollars, and it would be pretty much all you could drink—beer and Jell-O shots and other things."

Come Friday night, the ground around the house would shake from the intensity of the bass being pumped out by Ressi's speakers. Maye visited one of the parties and discovered that Ressi had hammered objects into the walls and lacquered them with glow-in-the-dark paint. She ended up working the door as the coat check/money taker and grabbed a pair of scissors for protection as the cash piled up in a shoe box.

A second house had fourteen rooms. Musk, Ressi, and one other person lived there. They fashioned tables by laying plywood on top of used kegs and came up with other makeshift furniture ideas. Musk returned home one day to find that Ressi had nailed his desk to the wall and then painted it in Day-Glo colors. Musk retaliated by pulling his desk down, painting it black, and studying. "I'm like, 'Dude, that's installation art in our party house,'" said Ressi. Remind Musk of this incident and he'll respond matter-of-factly, "It was a desk."

Musk will have the occasional vodka and Diet Coke, but he's

not a big drinker and does not really care for the taste of alcohol. "Somebody had to stay sober during these parties," Musk said. "I was paying my own way through college and could make an entire month's rent in one night. Adeo was in charge of doing cool shit around the house, and I would run the party." As Ressi put it, "Elon was the most straight-laced dude you have ever met. He never drank. He never did anything. Zero. Literally nothing." The only time Ressi had to step in and moderate Musk's behavior came during video game binges that could go on for days.

Musk's longtime interest in solar power and in finding other new ways to harness energy expanded at Penn. In December 1994, he had to come up with a business plan for one of his classes and ended up writing a paper titled "The Importance of Being Solar." The document started with a bit of Musk's wry sense of humor. At the top of the page, he wrote: "The sun will come out tomorrow. . . ."—Little Orphan Annie on the subject of renewable energy. The paper went on to predict a rise in solar power technology based on materials improvements and the construction of large-scale solar plants. Musk delved deeply into how solar cells work and the various compounds that can make them more efficient. He concluded the paper with a drawing of the "power station of the future." It depicted a pair of giant solar arrays in space—each four kilometers in width—sending their juice down to Earth via microwave beams to a receiving antenna with a seven-kilometer diameter. Musk received a 98 on what his professor deemed a "very interesting and well written paper."

A second paper talked about taking research documents and books and electronically scanning them, performing optical character recognition, and putting all of the information in a single database—much like a mix between today's Google Books and Google Scholar. And a third paper dwelled on another of Musk's favorite topics—ultracapacitors. In the forty-four-page document, Musk is plainly jubilant over the idea of a new form of energy storage that would suit his future pursuits with cars, planes, and rockets. Pointing to the latest research coming out of a lab in Silicon Valley, he wrote: "The end result represents the first new means of storing significant amounts of electrical energy since the development of the battery and fuel cell. Furthermore, because the Ultracapacitor retains the basic properties of a capacitor, it can deliver its energy over one hundred times faster than a battery of equivalent weight, and be recharged just as quickly." Musk received a 97 for this effort and praise for "a very thorough analysis" with "excellent financials!"

The remarks from the professor were spot-on. Musk's clear, concise writing is the work of a logician, moving from one point to the next with precision. What truly stood out, though, was Musk's ability to master difficult physics concepts in the midst of actual business plans. Even then, he showed an unusual knack for being able to perceive a path from a scientific advance to a for-profit enterprise.

As Musk began to think more seriously about what he would do after college, he briefly considered getting into the video-game business. He'd been obsessed with video games since his childhood and had held a gaming internship. But he came to see them as not quite grand enough a pursuit. "I really like computer games, but then if I made really great computer games, how much effect would that have on the world," he said. "It wouldn't have a big effect. Even though I have an intrinsic love of video games, I couldn't bring myself to do that as a career."

In interviews, Musk often makes sure that people know he had some truly big ideas on his mind during this period of his life. As he tells it, he would daydream at Queen's and Penn and usually end up with the same conclusion: he viewed the Internet, renewable energy, and space as the three areas that would undergo sig-

nificant change in the years to come and as the markets where he could make a big impact. He vowed to pursue projects in all three. "I told all my ex-girlfriends and my ex-wife about these ideas," he said. "It probably sounded like super-crazy talk."

Musk's insistence on explaining the early origins of his passion for electric cars, solar energy, and rockets can come off as insecure. It feels as if Musk is trying to shape his life story in a forced way. But for Musk, the distinction between stumbling into something and having intent is important. Musk has long wanted the world to know that he's different from the run-of-the-mill entrepreneur in Silicon Valley. He wasn't just sniffing out trends, and he wasn't consumed by the idea of getting rich. He's been in pursuit of a master plan all along. "I really was thinking about this stuff in college," he said. "It is not some invented story after the fact. I don't want to seem like a Johnny-come-lately or that I'm chasing a fad or just being opportunistic. I'm not an investor. I like to make technologies real that I think are important for the future and useful in some sort of way."

4

ELON'S FIRST START-UP

IN THE SUMMER OF 1994, MUSK AND HIS BROTHER, KIMBAL, TOOK THEIR first steps toward becoming honest-to-God Americans. They set off on a road trip across the country.

Kimbal had been working as a franchisee for College Pro Painters and done well for himself, running what amounted to a small business. He sold off his part of the franchise and pooled the money with what Musk had on hand to buy a beat-up 1970s BMW 320i. The brothers began their trip near San Francisco in August, as temperatures in California soared. The first part of the drive took them down to Needles, a city in the Mojave Desert. There they experienced the sweaty thrill of 120-degree weather in a car with no air-conditioning and learned to love pit stops at Carl's Jr. burger joints, where they spent hours recuperating in the cold.

The trip provided plenty of time for your typical twenty-something hijinks and raging capitalist daydreaming. The Web had just started to become accessible to the public thanks to the rise of directory sites like Yahoo! and tools like Netscape's browser. The brothers were tuned in to the Internet and thought they might like to start a company together doing something on the Web. From California to Colorado, Wyoming, South Dakota, and Illinois, they took turns driving, brainstorming, and talking shit before heading back east to get Musk to school that fall. The best idea to arise from the journey was an online network for doctors. This wasn't meant to be something as ambitious as electronic health records but more of a system for physicians to exchange information and collaborate. "It seemed like the medical industry was one that could be disrupted," Kimbal said. "I went to work on a business plan and the sales and marketing side of it later, but it didn't fly. We didn't love it."

Musk had spent the earlier part of that summer in Silicon Valley, holding down a pair of internships. By day, he worked at Pinnacle Research Institute. Based in Los Gatos, Pinnacle was a much-ballyhooed start-up with a team of scientists exploring ways in which ultracapacitors could be used as a revolutionary fuel source in electric and hybrid vehicles. The work also veered—at least conceptually—into more bizarre territory. Musk could talk at length about how ultracapacitors might be used to build laser-based sidearms in the tradition of *Star Wars* and just about any other futuristic film. The laser guns would release rounds of enormous energy, and then the shooter would replace an ultra-capacitor at the base of the gun, much like swapping out a clip of bullets, and start blasting away again. Ultracapacitors also looked promising as the power supplies for missiles. They were more resilient than batteries under the mechanical stresses of a launch and would hold a more consistent charge over long periods of time. Musk fell in love with the work at Pinnacle and began using it as the basis for some of his business plan experiments at Penn and for his industrialist fantasies.

In the evenings, Musk headed to Rocket Science Games, a start-up based in Palo Alto that wanted to create the most advanced video games ever made by moving them off cartridges and onto CDs that could hold more information. The CDs would in theory allow them to bring Hollywood-style storytelling and production quality to the games. A team of budding all-stars who were a mix of engineers and film people was assembled to pull off the work. Tony Fadell, who would later drive much of the development of both the iPod and iPhone at Apple, worked at Rocket Science, as did the guys who developed the QuickTime multimedia software for Apple. They also had people who worked on the original *Star Wars* effects at Industrial Light & Magic and some who did games at LucasArts Entertainment. Rocket Science gave Musk a flavor for what Silicon Valley had to offer both from a talent and culture perspective. There were people working at the office twenty-four hours a day, and they didn't think it at all odd that Musk would turn up around 5 p.m. every evening to start his second job. "We brought him in to write some very menial low-level code," said Peter Barrett, an Australian engineer who helped start the company. "He was completely unflappable. After a short while, I don't think anyone was giving him any direction, and he ended up making what he wanted to make."

Specifically, Musk had been asked to write the drivers that would let joysticks and mice communicate with various computers and games. Drivers are the same types of annoying files that you have to install to get a printer or camera working with a home computer—true grunt work. A self-taught programmer, Musk fancied himself quite good at coding and assigned himself to more ambitious jobs. "I was basically trying to figure out how you could multitask stuff, so you could read video from a CD, while running a game at the same time," Musk said. "At the time, you could do one or the other. It was this complicated bit of assembly programming." Complicated indeed. Musk had to issue commands that spoke directly to a computer's main microprocessor and fiddled with the most basic functions that made the machine work. Bruce Leak, the former lead engineer behind Apple's QuickTime, had overseen the hiring of Musk and marveled at his ability to pull all-nighters. "He had boundless energy," Leak said. "Kids these days have no idea about hardware or how stuff works, but he had a PC hacker background and was not afraid to just go figure things out."

Musk found in Silicon Valley a wealth of the opportunity he'd been seeking and a place equal to his ambitions. He would return two summers in a row and then bolt west permanently after graduating with dual degrees from Penn. He initially intended to pursue a doctorate in materials science and physics at Stanford and to advance the work he'd done at Pinnacle on ultracapacitors. As the story goes, Musk dropped out of Stanford after two days, finding the Internet's call irresistible. He talked Kimbal into moving to Silicon Valley as well, so they could conquer the Web together.

The first inklings of a viable Internet business had come to Musk during his internships. A salesperson from the Yellow Pages had come into one of the start-up offices. He tried to sell the idea of an online listing to complement the regular listing a company would have in the big, fat Yellow Pages book. The salesman struggled with his pitch and clearly had little grasp of what the Internet actually was or how someone would find a business on it. The flimsy pitch got Musk thinking, and he reached out to Kimbal, talking up the idea of helping businesses get online for the first time.

"Elon said, 'These guys don't know what they are talking about. Maybe this is something we can do,'" Kimbal said. This was 1995, and the brothers were about to form Global Link Information Network, a start-up that would eventually be renamed

Zip2. (For details on the controversy surrounding Zip2's founding and Musk's academic record, see Appendix 1.)

The Zip2 idea was ingenious. Few small businesses in 1995
understood the ramifications of the Internet. They had little idea
how to get on it and didn't really see the value in creating a website for their business or even in having a Yellow Pages–like listing online. Musk and his brother hoped to convince restaurants,
clothing shops, hairdressers, and the like that the time had come
for them to make their presence known to the Web-surfing public. Zip2 would create a searchable directory of businesses and tie
this into maps. Musk often explained the concept through pizza,
saying that everyone deserved the right to know the location of
their closest pizza parlor and the turn-by-turn directions to get
there. This may seem obvious today—think Yelp meets Google
Maps—but back then, not even stoners had dreamed up such a
service.

The Musk brothers brought Zip2 to life at 430 Sherman Avenue in Palo Alto. They rented a studio-apartment-sized office—
twenty feet by thirty feet—and acquired some basic furniture.
The three-story building had its quirks. There were no elevators,
and the toilets often backed up. "It was literally a shitty place to
work," said an early employee. To get a fast Internet connection,
Musk struck a deal with Ray Girouard, an entrepreneur who ran
an Internet service provider operation from the floor below the
Zip2 offices. According to Girouard, Musk drilled a hole in the
drywall near the Zip2 door and then strung an Ethernet cable
down the stairwell to the ISP. "They were slow to pay a couple of
times but never stiffed me on the bill," Girouard said.

Musk did all of the original coding behind the service himself, while the more amiable Kimbal looked to ramp up the door-
to-door sales operation. Musk had acquired a cheap license to
a database of business listings in the Bay Area that would give

a business's name and its address. He then contacted Navteq, a company that had spent hundreds of millions of dollars to create digital maps and directions that could be used in early GPS navigation-style devices, and struck a masterful bargain. "We called them up, and they gave us the technology for free," said Kimbal. Musk merged the two databases together to get a rudimentary system up and running. Over time, Zip2's engineers had to augment this initial data haul with more maps to cover areas outside of major metropolitan areas and to build custom turn-by-turn directions that would look good and work well on a home computer.

Errol Musk gave his sons $28,000 to help them through this period, but they were more or less broke after getting the office space, licensing software, and buying some equipment. For the first three months of Zip2's life, Musk and his brother lived at the office. They had a small closet where they kept their clothes and would shower at the YMCA. "Sometimes we ate four meals a day at Jack in the Box," Kimbal said. "It was open twenty-four hours, which suited our work schedule. I got a smoothie one time, and there was something in it. I just pulled it out and kept drinking. I haven't been able to eat there since, but I can still recite their menu."

Next, the brothers rented a two-bedroom apartment. They didn't have the money or the inclination to get furniture. So there were just a couple of mattresses on the floor. Musk somehow managed to convince a young South Korean engineer to come work at Zip2 as an intern in exchange for room and board. "This poor kid thought he was coming over for a job at a big company," Kimbal said. "He ended up living with us and had no idea what he was getting into." One day, the intern drove the Musks' battered BMW 320i to work, and a wheel came off en route. The axle dug into the street at the intersection of Page Mill Road and

El Camino Real, and the groove it carved out remained visible for years.

Zip2 may have been a go-go Internet enterprise aimed at the Information Age, but getting it off the ground required old-fashioned door-to-door salesmanship. Businesses needed to be persuaded of the Web's benefits and charmed into paying for the unknown. In late 1995, the Musk brothers began making their first hires and assembling a motley sales team. Jeff Heilman, a free-spirited twenty-year-old trying to figure out what to do with his life, arrived as one of Zip2's first recruits. He'd been watching TV late one night with his dad and seen a Web address printed at the bottom of the screen during a commercial. "It was for something dot-com," Heilman said. "I remember sitting there and asking my dad what we were looking at. He said he didn't know, either. That's when I realized I had to go find me some Internet." Heilman spent a couple of weeks trying to chat up people who could explain the Internet to him and then stumbled on a two-by-two-inch Zip2 job listing in the *San Jose Mercury News*. "Internet Sales Apply Here!" it read, and Heilman got the gig. A handful of other salespeople joined him and worked for commissions.

Musk never seemed to leave the office. He slept, not unlike a dog, on a beanbag next to his desk. "Almost every day, I'd come in at seven thirty or eight A.M., and he'd be asleep right there on that bag," Heilman said. "Maybe he showered on the weekends. I don't know." Musk asked those first employees of Zip2 to give him a kick when they arrived, and he'd wake up and get back to work. While Musk did his possessed coder thing, Kimbal became the rah-rah sales leader. "Kimbal was the eternal optimist, and he was very, very uplifting," Heilman said. "I had never met anyone quite like him." Kimbal sent Heilman to the high-end Stanford shopping mall and to University Avenue, the main drag in Palo Alto, to coax retailers into signing up with Zip2, explaining that a sponsored listing would send a company to the top of search results. The big problem, of course, was that no one was buying. Week after week, Heilman knocked on doors and returned to the office with very little to report in the way of good news. The nicest responses came from the people who told Heilman that advertising on the Internet sounded like the dumbest thing they had ever heard of. Most often, the shop owners just told Heilman to leave and stop bothering them. When lunchtime came around, the Musks would reach into a cigar box where they kept some cash, take Heilman out, and get the depressing status reports on the sales.

Craig Mohr, another early employee, gave up his job selling real estate to hawk Zip2's service. He decided to court auto dealerships because they usually spent lots of money on advertising. He told them about Zip2's main website—www.totalinfo.com— and tried to convince them that demand was high to get a listing like www.totalinfo.com/toyotaofsiliconvalley. The service did not always work when Mohr demonstrated it or it would load very slowly, as was common back then. This forced him to talk the customers into imagining Zip2's potential. "One day I came back with about nine hundred dollars in checks," Mohr said. "I walked into the office and asked the guys what they wanted me to do with the money. Elon stopped pounding his keyboard, leaned out from behind his monitor, and said, 'No way, you've got money.'"

What kept the employees' spirits up were the continuous improvements Musk made with the Zip2 software. The service had morphed from a proof-of-concept to an actual product that could be used and demoed. Ever marketing savvy, the Musk brothers tried to make their Web service seem more important by giving it an imposing physical body. Musk built a huge case around a standard PC and lugged the unit onto a base with wheels. When prospective investors would come by, Musk would put on a show and roll this massive machine out so that it appeared like Zip2 ran inside of a mini-supercomputer. "The investors thought that was impressive," Kimbal said. Heilman also noticed that the investors bought into Musk's slavish devotion to the company. "Even then, as essentially a college kid with zits, Elon had this drive that this thing—whatever it was—had to get done and that if he didn't do it, he'd miss his shot," Heilman said. "I think that's what the VCs saw—that he was willing to stake his existence on building out this platform." Musk actually said as much to one venture capitalist, informing him, "My mentality is that of a samurai. I would rather commit seppuku than fail."

Early on in the Zip2 venture, Musk acquired an important confidant, who tempered some of these more dramatic impulses. Greg Kouri, a Canadian businessman in his mid-thirties, had met the Musks in Toronto and bought into the early Zip2 brainstorming. The boys had showed up at his door one morning to inform Kouri that they intended to head to California to give the business a shot. Still in his red bathrobe, Kouri went back into the house, dug around for a couple of minutes, and came back with a wad of $6,000. In early 1996, he moved to California and joined Zip2 as a cofounder.

Kouri, who had done a number of real estate deals in the past and had actual business experience and skills at reading people, served as the adult supervision at Zip2. The Canadian had a knack for calming Musk and ended up becoming something of a mentor. "Really smart people sometimes don't understand that not everyone can keep up with them or go as fast," said Derek Proudian, a venture capitalist who would become Zip2's chief executive officer. "Greg is one of the few people that Elon would listen to and had a way of putting things in context for him." Kouri also used to referee fistfights between Elon and Kimbal, in the middle of the office.

"I don't get in fights with anyone else, but Elon and I don't have the ability to reconcile a vision other than our own," Kimbal said. During a particularly nasty scrap over a business decision, Elon ripped some skin off his fist and had to go get a tetanus shot. Kouri put an end to the fights after that. (Kouri died of a heart attack in 2012 at the age of fifty-one, having made a fortune investing in Musk's companies. Musk attended his funeral. "We owe him a lot," said Kimbal.)

In early 1996, Zip2 underwent a massive change. The venture capital firm Mohr Davidow Ventures had caught wind of a couple of South African boys trying to make a Yellow Pages for the Internet and met with the brothers. Musk, while raw in his presentation skills, pitched the company well enough, and the investors came away impressed with his energy. Mohr Davidow invested $3 million into the company.* With these funds in hand, the company officially changed its name from Global Link to Zip2—the idea being zip to here, zip to there—moved to a larger office at 390 Cambridge Avenue in Palo Alto, and began hiring talented engineers. Zip2 also shifted its business strategy. At the time, the company had built one of the best direction systems on the Web. Zip2 would advance this technology and take it from focusing just on the Bay Area to having a national scope. The company's main focus, however, would be an altogether new play. Instead of selling its service door-to-door, Zip2 would create a software package that could be sold to newspapers, which would in turn build their own directories for real estate, auto dealers,

---

* The Musk brothers were not the most aggressive businessmen at this point. "I remember from their business plan that they were originally asking for a ten-thousand-dollar investment for twenty-five percent of their company," said Steve Jurvetson, the venture capitalist. "That is a cheap deal! When I heard about the three-million-dollar investment, I wondered if Mohr Davidow had actually read the business plan. Somehow, the brothers ended up raising a normal venture round."

and classifieds. The newspapers were late understanding how the Internet would impact their businesses, and Zip2's software would give them a quick way of getting online without needing to develop all their own technology from scratch. For its part, Zip2 could chase bigger prey and get a cut of a nationwide network of listings.

This transition of the business model and the company's makeup would be a seminal moment in Musk's life. The venture capitalists pushed Musk into the role of chief technology officer and hired Rich Sorkin as the company's CEO. Sorkin had worked at Creative Labs, a maker of audio equipment, and run the business development group at the company, where he steered a number of investments in Internet start-ups. Zip2's investors saw him as experienced and clued in to the Web. While Musk agreed to the arrangement, he came to resent giving up control of Zip2. "Probably the biggest regret the whole time I worked with him was that he had made a deal with the devil with Mohr Davidow," said Jim Ambras, the vice president of engineering at Zip2. "Elon didn't have any operational responsibilities, and he wanted to be CEO."

Ambras had worked at Hewlett-Packard Labs and Silicon Graphics Inc. and exemplified the high-caliber talent Zip2 brought on after the first wave of money arrived. Silicon Graphics, a maker of high-end computers beloved by Hollywood, was the flashiest company of its day and had hoarded the elite geeks of Silicon Valley. And yet Ambras used the promise of Internet riches to poach a team of SGI's smartest engineers over to Zip2. "Our attorneys got a letter from SGI saying that we were cherry-picking the very best guys," Ambras said. "Elon thought that was fantastic."

While Musk had exceled as a self-taught coder, his skills weren't nearly as polished as those of the new hires. They took one look at Zip2's code and began rewriting the vast majority of the software. Musk bristled at some of their changes, but the computer scientists needed just a fraction of the lines of code that Musk used to get their jobs done. They had a knack for dividing software projects into chunks that could be altered and refined whereas Musk fell into the classic self-taught coder trap of writing what developers call hairballs—big, monolithic hunks of code that could go berserk for mysterious reasons. The engineers also brought a more refined working structure and realistic deadlines to the engineering group. This was a welcome change from Musk's approach, which had been to set overly optimistic deadlines and then try to get engineers to work nonstop for days on end to meet the goals. "If you asked Elon how long it would take to do something, there was never anything in his mind that would take more than an hour," Ambras said. "We came to interpret an hour as really taking a day or two and if Elon ever did say something would take a day, we allowed for a week or two weeks."

Starting Zip2 and watching it grow imbued Musk with self-confidence. Terence Beney, one of Musk's high school friends, came to California for a visit and noticed the change in Musk's character right away. He watched Musk confront a nasty landlord who had been giving his mother, who was renting an apartment in town, a hard time. "He said, 'If you're going to bully someone, bully me.' It was startling to see him take over the situation. The last time I had seen him he was this geeky, awkward kid who would sometimes lose his temper. He was the kid you would pick on to get a response. Now he was confident and in control." Musk also began consciously trying to manage his criticism of others. "Elon is not someone who would say, 'I feel you. I see your point of view,'" said Justine. "Because he doesn't have that 'I feel you' dimension there were things that seemed obvious to other people that weren't that obvious to him. He had to learn that a twenty-something-year-old shouldn't really shoot down the plans of older, senior people and point out everything wrong with them. He learned to modify his behavior in certain ways. I just think he comes at the world through strategy and intellect." The personality tweaks worked with varying degrees of success. Musk still tended to drive the young engineers mad with his work demands and blunt criticism. "I remember being in a meeting once brainstorming about a new product—a new-car site," said Doris Downes, the creative director at Zip2. "Someone complained about a technical change that we wanted being impossible. Elon turned and said, 'I don't really give a damn what you think,' and walked out of the meeting. For Elon, the word *no* does not exist, and he expects that attitude from everyone around him." Periodically, Musk let loose on the more senior executives as well. "You would see people come out of the meetings with this disgusted look on their face," Mohr, the salesman, said. "You don't get to where Elon is now by always being a nice guy, and he was just so driven and sure of himself."

As Musk tried to come to terms with the changes the investors had inflicted on Zip2, he did enjoy some of the perks of having big-money backing. The financiers helped the Musk brothers with their visas. They also gave them $30,000 each to buy cars. Musk and Kimbal had traded in their dilapidated BMW for a dilapidated sedan that they spray-painted with polka dots. Kimbal upgraded from that to a BMW 3 Series, and Musk bought a Jaguar E-Type. "It kept breaking down, and would arrive at the office on a flatbed," Kimbal said. "But Elon always thought big."*

---

* Musk also got to show off the new office to his mother, Maye, and Justine. Maye sometimes sat in on meetings and came up with the idea of adding a "reverse directions" button on the Zip2 maps, which let people flip around their journeys and ended up becoming a popular feature on all mapping services.

As a bonding exercise one weekend, Musk, Ambras, a few other employees and friends took off for a bike ride through the Saratoga Gap trail in the Santa Cruz Mountains. Most of the riders had been training and were accustomed to strenuous sessions and the summer's heat. They set up the mountains at a furious pace. After an hour, Russ Rive, Musk's cousin, reached the top and proceeded to vomit. Right behind him were the rest of the cyclists. Then, fifteen minutes later, Musk became visible to the group. His face had turned purple, and sweat poured out of him, and he made it to the top. "I always think back to that ride. He wasn't close to being in the condition needed for it," Ambras said. "Anyone else would have quit or walked up their bike. As I watched him climb that final hundred feet with suffering all over his face, I thought, That's Elon. Do or die but don't give up."

Musk continued to be a ball of energy around the office as well. Ahead of visits by venture capitalists and other investors, Musk would rally the troops and instruct them all to get on the phone to create a buzzy atmosphere. He also formed a video-game team to participate in competitions around Quake, a first-person-shooter game. "We competed in one of the first nationwide tournaments," Musk said. "We came in second, and we would have come in first, but one of our top players' machine crashed because he had pushed his graphics card too hard. We won a few thousand dollars."

Zip2 had remarkable success courting newspapers. The *New York Times*, Knight Ridder, Hearst Corporation, and other media properties signed up to its service. Some of these companies contributed $50 million in additional funding for Zip2. Services like Craigslist with its free online classifieds had just started to appear, and the newspapers needed some course of action. "The newspapers knew they were in trouble with the Internet, and the idea was to sign up as many of them as possible," Ambras said.

"They wanted classifieds and listings for real estate, automotive, and entertainment and could use us as a platform for all these online services." Zip2 acquired a trademark for its "We Power the Press" slogan and the influx of cash kept Zip2 growing fast. Company headquarters were soon so crowded that one desk ended up directly in front of the women's bathroom. In 1997, Zip2 moved into flashier, more spacious digs at 444 Castro Street in Mountain View.

It irritated Musk that Zip2 had become a behind-the-scenes player to the newspapers. He believed the company could offer interesting services directly to consumers and encouraged the purchase of the domain name city.com with the hopes of turning it into a consumer destination. But the lure of the media companies' money kept Sorkin and the board on a conservative path, and they decided to worry about a consumer push down the road.

In April 1998, Zip2 announced a blockbuster move to double down on its strategy. It would merge with its main competitor CitySearch in a deal valued at around $300 million. The new company would retain the CitySearch name, while Sorkin would head up the venture. On paper, the union looked very much like a merger of equals. CitySearch had built up an extensive set of directories for cities around the country. It also appeared to have strong sales and marketing teams that would complement the talented engineers at Zip2. The merger had been announced in the press and seemed inevitable.

The opinions on what happened next vary greatly. The logistics of the situation required the two companies to go over each other's books and to figure out which employees would be fired to avoid a duplication of roles. This process raised some questions about how frank CitySearch had been with its financials and rankled some executives at Zip2 who could see their positions being diminished or erased altogether at the new company. One faction inside Zip2 argued that the deal should be abandoned, while Sorkin demanded that it go through. Musk, who had been an early advocate of the deal, turned against it. In May 1998, the two companies canceled the merger, and the press pounced, making a big deal of the chaotic bust-up. Musk urged Zip2's board to oust Sorkin and reinstate him as CEO of Zip2. The board declined. Instead, Musk lost his chairman title, and Sorkin was replaced by Derek Proudian, a venture capitalist with Mohr Davidow. Sorkin considered Musk's behavior through the whole affair atrocious and later pointed to the board's reaction and Musk's demotion as evidence that they felt the same way. "There was a lot of backlash and finger-pointing," Proudian said. "Elon wanted to be CEO, but I said, 'This is your first company. Let's find an acquirer and make some money, so you can do your second, third, and fourth company.'"

With the deal busted, Zip2 found itself in a predicament. It was losing money. Musk still wanted to go the consumer route, but Proudian feared that would take too much capital. Microsoft had mounted a charge into the same market, and start-ups with mapping, real estate, and automotive ideas multiplied. The Zip2 engineers were deflated and worried that they might not be able to outrun the competition. Then, in February 1999, the PC maker Compaq Computer suddenly offered to pay $307 million in cash for Zip2. "It was like pennies from heaven," said Ed Ho, a former Zip2 executive. Zip2's board accepted the offer, and the company rented out a restaurant in Palo Alto and threw a huge party. Mohr Davidow had made back twenty times its original investment, and Musk and Kimbal had come away with $22 million and $15 million, respectively. Musk never entertained the idea of sticking around at Compaq. "As soon as it was clear the company would be sold, Elon was on to his next project," Proudian said. From that point on, Musk would fight to maintain control of his companies and stay CEO. "We were overwhelmed and just thought these guys must know what they're doing," Kimbal said. "But they' didn't. There was no vision once they took over. They were investors, and we got on well with them, but the vision had just disappeared from the company."

Years later, after he had time to reflect on the Zip2 situation, Musk realized that he could have handled some of the situations with employees better. "I had never really run a team of any sort before," Musk said. "I'd never been a sports captain or a captain of anything or managed a single person. I had to think, Okay, what are the things that affect how a team functions. The first obvious assumption would be that other people will behave like you. But that's not true. Even if they would like to behave like you, they don't necessarily have all the assumptions or information that you have in your mind. So, if I know a certain set of things, and I talk to a replica of myself but only communicate half the information, you can't expect that the replica would come to the same conclusion. You have to put yourself in a position where you say, 'Well, how would this sound to them, knowing what they know?'"

Employees at Zip2 would go home at night, come back, and find that Musk had changed their work without talking to them, and Musk's confrontational style did more harm than good. "Yeah, we had some very good software engineers at Zip2, but I mean, I could code way better than them. And I'd just go in and fix their fucking code," Musk said. "I would be frustrated waiting for their stuff, so I'm going to go and fix your code and now it runs five times faster, you idiot. There was one guy who wrote a quantum mechanics equation, a quantum probability on the board, and he got it wrong. I'm like, 'How can you write that?' Then I corrected it for him. He hated me after that. Eventually, I realized, Okay, I might have fixed that thing but now I've made the person unproductive. It just wasn't a good way to go about things."

Musk, the dot-com striver, had been both lucky and good. He had a decent idea, turned it into a real service, and came out of the dot-com tumult with cash in his pockets, which was better than what many of his compatriots could say. The process had been painful. Musk had yearned to be a leader, but the people around him struggled to see how Musk as the CEO could work. As far as Musk was concerned, they were all wrong, and he set out to prove his point with what would end up being even more dramatic results.

5

PAYPAL MAFIA BOSS

T HE SALE OF ZIP2 INFUSED ELON MUSK WITH A NEW BRAND OF CONFI-
dence. Much like the video-game characters he adored, Musk
had leveled up. He had solved Silicon Valley and become
what everyone at the time wanted to be—a dot-com millionaire.
His next venture would need to live up to his rapidly inflating
ambition. This left Musk searching for an industry that had tons
of money and inefficiencies that he and the Internet could exploit.
Musk began thinking back to his time as an intern at the Bank
of Nova Scotia. His big takeaway from that job, that bankers are
rich and dumb, now had the feel of a massive opportunity.

During his time working for the head of strategy at the bank
in the early 1990s, Musk had been asked to take a look at the company's third-world debt portfolio. This pool of money went by
the depressing name of "less-developed country debt," and Bank
of Nova Scotia had billions of dollars of it. Countries throughout
South America and elsewhere had defaulted in the years prior,
forcing the bank to write down some of its debt value. Musk's

boss wanted him to dig into the bank's holdings as a learning experiment and try to determine how much the debt was actually worth.

While pursuing this project, Musk stumbled upon what seemed like an obvious business opportunity. The United States had tried to help reduce the debt burden of a number of developing countries through so-called Brady bonds, in which the U.S. government basically backstopped the debt of countries like Brazil and Argentina. Musk noticed an arbitrage play. "I calculated the backstop value, and it was something like fifty cents on the dollar, while the actual debt was trading at twenty-five cents," Musk said. "This was like the biggest opportunity ever, and nobody seemed to realize it." Musk tried to remain cool and calm as he rang Goldman Sachs, one of the main traders in this market, and probed around about what he had seen. He inquired as to how much Brazilian debt might be available at the 25-cents price. "The guy said, 'How much do you want?' and I came up with some ridiculous number like ten billion dollars," Musk said. When the trader confirmed that was doable, Musk hung up the phone. "I was thinking that they had to be fucking crazy because you could double your money. Everything was backed by Uncle Sam. It was a no-brainer."

Musk had spent the summer earning about fourteen dollars an hour and getting chewed out for using the executive coffee machine, among other status infractions, and figured his moment to shine and make a big bonus had arrived. He sprinted up to his boss's office and pitched the opportunity of a lifetime. "You can make billions of dollars for free," he said. His boss told Musk to write up a report, which soon got passed up to the bank's CEO, who promptly rejected the proposal, saying the bank had been burned on Brazilian and Argentinian debt before and didn't want to mess with it again. "I tried to tell them that's not the point,"

Musk said. "The point is that it's fucking backed by Uncle Sam. It doesn't matter what the South Americans do. You cannot lose unless you think the U.S. Treasury is going to default. But they still didn't do it, and I was stunned. Later in life, as I competed against the banks, I would think back to this moment, and it gave me confidence. All the bankers did was copy what everyone else did. If everyone else ran off a bloody cliff, they'd run right off a cliff with them. If there was a giant pile of gold sitting in the middle of the room and nobody was picking it up, they wouldn't pick it up, either."

In the years that followed, Musk considered starting an Internet bank and discussed it openly during his internship at Pinnacle Research in 1995. The youthful Musk lectured the scientists about the inevitable transition coming in finance toward online systems, but they tried to talk him down, saying that it would takes ages for Web security to be good enough to win over consumers. Musk, though, remained convinced that the finance industry could do with a major upgrade and that he could have a big influence on banking with a relatively small investment. "Money is low bandwidth," he said, during a speech at Stanford University in 2003, to describe his thinking. "You don't need some sort of big infrastructure improvement to do things with it. It's really just an entry in a database."

The actual plan that Musk concocted was beyond grandiose. As the researchers at Pinnacle had pointed out, people were barely comfortable buying books online. They might take their chances entering a credit card number but exposing just their bank accounts to the Web was out of the question to many. Pah. So what? Musk wanted to build a full-service financial institution online: a company that would have savings and checking accounts as well as brokerage services and insurance. The technology to build such a service was possible, but navigating the regulatory hell of creating an online bank from scratch looked like an intractable problem to optimists and an impossibility to more level heads. This was not dishing out directions to a pizzeria or putting up a house listing. It was dealing with people's finances, and there would be real repercussions if the service did not work as billed.

Undaunted, Musk kicked this new plan into action before Zip2 had even been sold. He chatted up some of the best engineers at the company to get a feel for who might be willing to join him in another venture. Musk also bounced his ideas off some contacts he'd made at the bank in Canada. In January 1999, with Zip2's board seeking a buyer, Musk began to formalize his banking plan. The deal with Compaq was announced the next month. And in March, Musk incorporated X.com, a finance start-up with a pornographic-sounding name.

It had taken Musk less than a decade to go from being a Canadian backpacker to becoming a multimillionaire at the age of twenty-seven. With his $22 million, he moved from sharing an apartment with three roommates to buying an 1,800-square-foot condo and renovating it. He also bought a $1 million McLaren F1 sports car and a small prop plane and learned to fly. Musk embraced the newfound celebrity that he'd earned as part of the dot-com millionaire set. He let CNN show up at his apartment at 7 A.M. to film the delivery of the car. A black eighteen-wheeler pulled up in front of Musk's place and then lowered the sleek, sliver vehicle onto the street, while Musk stood slack-jawed with his arms folded. "There are sixty-two McLarens in the world, and I will own one of them," he told CNN. "Wow, I can't believe it's actually here. That's pretty wild, man."

CNN interspersed video of the car delivery with interviews with Musk. The whole time he looked like a caricature of an engineer who had made it big. Musk's hair had started thinning, and he had a closely cropped cut that accentuated his boyish face. He wore an all-too-big brown sport coat and checked his cell phone from his lavish car, sitting next to his gorgeous girlfriend, Justine, and he seemed spellbound by his life. Musk rolled out one laughable rich-guy line after another, talking first about the Zip2 deal—"Receiving cash is cash. I mean, those are just a large number of Ben Franklins"—next about the awesomeness of his life—"There it is, gentlemen, the fastest car in the world"—and then about his prodigious ambition—"I could go and buy one of the islands in the Bahamas and turn it into my personal fiefdom, but I am much more interested in trying to build and create a new company." The camera crew followed Musk to the X.com offices, where his cocksure delivery led to another round of cringe-worthy statements: "I do not fit the picture of a banker," "Raising fifty million dollars is a matter of making a series of phone calls, and the money is there," "I think X.com could absolutely be a multibillion-dollar bonanza."

Musk purchased the McLaren from a seller in Florida, snatching the car away from Ralph Lauren, who had also inquired about buying it. Even very wealthy people like Lauren would tend to reserve something like a McLaren for special events or the occasional Sunday drive. Not Musk. He drove it all around Silicon Valley and parked it on the street by the X.com offices. His friends were horrified to see such a work of art covered with bird droppings or in the parking lot of a Safeway. One day, Musk e-mailed fellow McLaren owner Larry Ellison, the billionaire cofounder of the software maker Oracle, out of the blue to see if he wanted to go race cars around a track for fun. Jim Clark, another billionaire who liked fast things, caught wind of the proposal and told a friend that he needed to rush over to the local Ferrari dealership to buy something that could compete. Musk had joined the big boys' club. "Elon was super-excited about all of this," said George Zachary, a venture capitalist and close friend of Musk's. "He showed me the correspondence with Larry." The next year, while driving down Sand Hill Road to meet with an investor, Musk turned to a friend in the car and said, "Watch this." He floored the car, did a lane change, spun out, and hit an embankment, which started the car spinning in midair like a Frisbee. The windows and wheels were blown to smithereens, and the body of the car damaged. Musk again turned to his companion and said, "The funny part is it wasn't insured." The two of them then thumbed a ride to the venture capitalist's office.

To his credit, Musk did not fully buy in to this playboy persona. He actually plowed the majority of the money he made from Zip2 into X.com. There were practical reasons for this decision. Investors catch a break under the tax law if they roll a windfall into a new venture within a couple of months. But even by Silicon Valley's high-risk standards, it was shocking to put so much of one's newfound wealth into something as iffy as an online bank. All told, Musk invested about $12 million into X.com, leaving him, after taxes, with $4 million or so for personal use. "That's part of what separates Elon from mere mortals," said Ed Ho, the former Zip2 executive, who went on to cofound X.com. "He's willing to take an insane amount of personal risk. When you do a deal like that, it either pays off or you end up in a bus shelter somewhere."

Musk's decision to invest so much money in X.com looks even more unusual in hindsight. Much of the point of being a dot-com success in 1999 was to prove yourself once, stash away your millions, and then use your credentials to talk other people into betting their money on your next venture. Musk would certainly go on to rely on outside investors, but he put major skin in the game as well. So while Musk could be found on television talking like the rest of the self-absorbed dot-com schmucks, he behaved more like a throwback to Silicon Valley's earlier days, when the founders of companies like Intel were willing to take huge gambles on themselves.

Where Zip2 had been a neat, useful idea, X.com held the promise of fomenting a major revolution. Musk, for the first time, would be confronting a deep-pocketed, entrenched industry head-on with the hopes of upending all of the incumbents. Musk also began to hone his trademark style of entering an ultracomplex business and not letting the fact that he knew very little about the industry's nuances bother him in the slightest. He had an inkling that the bankers were doing finance all wrong and that he could run the business better than everyone else. Musk's ego and confidence had started heading toward the levels that would inspire some and leave others thinking of him as pompous and unscrupulous. The creation of X.com would ultimately reveal a great deal about Musk's creativity, relentless drive, confrontational style, and foibles as a leader. Musk would also get another taste of being pushed aside at his own company and the pain that accompanies a grand vision left unfulfilled.

Musk assembled what looked like an all-star crew to start X.com. Ho had worked at SGI and Zip2 as an engineer, and his peers marveled at his coding and team-management skills. They were joined by a pair of Canadians with finance experience— Harris Fricker and Christopher Payne. Musk had met Fricker during his time as an intern at the Bank of Nova Scotia, and the two really hit it off. A Rhodes scholar, Fricker brought the knowledge of the banking world's mechanics that X.com would need. Payne was Fricker's friend from the Canadian finance community. All four men were considered cofounders of the company, while Musk emerged as the largest shareholder thanks to his hefty up-front investment. X.com began, like so many Silicon

Valley operations, at a house where the cofounders began brain-storming, and then moved to more formal offices at 394 University Avenue in Palo Alto.

The cofounders were aligned philosophically around the idea that the banking industry had fallen behind the times. Visiting a branch bank to speak with a teller seemed pretty archaic now that the Internet had arrived. The rhetoric sounded good, and the four men were enthused. The only thing stopping them was reality. Musk had a modicum of banking experience and had resorted to buying a book on the industry to help understand its inner workings. The more the cofounders thought about their plan of attack, the more they realized the regulatory issues blocking the creation of an online bank were insurmountable. "As four and five months went by, the onion just kept unwrapping," said Ho.*

From the outset, there were personality clashes as well. Musk had become a budding superstar in Silicon Valley and had the press fawning over him. This didn't sit that well with Fricker, who'd moved from Canada and pegged X.com as his chance to make a mark on the world as a banking whiz. Fricker, according to numerous people, wanted to run X.com and do so in a more conventional manner. He found Musk's visionary statements to the press about rethinking the entire banking industry silly since the company was struggling to build much of anything. "We were out promising the sun, moon, and the stars to the media," Fricker said. "Elon would say that this is not a normal business environment, and you have to suspend normal business think-ing. He said, 'There is a happy-gas factory up on the hill, and

---

* At one point, the founders thought the easiest way to solve their prob-lems would just be to buy a bank and revamp it. While that didn't happen, they did snag a high-profile controller from Bank of America, who in turn explained, in painful detail, the complexities of sourcing loans, transferring money, and protecting accounts.

it's pumping stuff into the Valley.'" Fricker would not be the last person to accuse Musk of overhyping products and playing the public, although whether this is a flaw or one of Musk's great talents as a businessman is up for debate.

The squabble between Fricker and Musk came to a quick, nasty end. Just five months after X.com had started, Fricker initiated a coup. "He said either he takes over as CEO or he's just going to take everyone from the company and create his own company," Musk said. "I don't do well with blackmail. I said, 'You should go do that.' So he did." Musk tried to talk Ho and some of the other key engineers into staying, but they sided with Fricker and left. Musk ended up with a shell of a company and a handful of loyal employees. "After all that went down, I remember sitting with Elon in his office," said Julie Ankenbrandt, an early X.com employee who stayed. "There were a million laws in place to block something like X.com from happening, but Elon didn't care. He just looked at me and said, 'I guess we should hire some more people.'"*

Musk had been trying to raise funding for X.com and had been forced to go to venture capitalists and confess that there wasn't much in the way of a company left. Mike Moritz, a famed investor from Sequoia Capital, backed the company nonethe-

---

* Fricker disputed that he yearned to be CEO, saying instead that the other employees had encouraged him to take over because of Musk's struggles getting the business off the ground. Fricker and Musk, once close friends, remain unimpressed with each other. "Elon has his own code of ethics and honor and plays the game extraordinarily hard," Fricker said. "When it comes down to it, for him, business is war." According to Musk, "Harris is very smart, but I don't think he has a good heart. He had a really intense desire to be running the show, and he wanted to take the company in ridiculous directions." Fricker went on to have a very successful career as CEO of GMP Capital, a Canadian financial services company. Payne founded a private equity firm in Toronto.

less, making a bet on Musk and little else. Musk hit the streets of Silicon Valley once again and managed to attract engineers with his rah-rah speeches about the future of Internet banking. Scott Anderson, a young computer scientist, started on August 1, 1999, just a few days after the exodus, and bought right into the vision. "You look back, and it was total insanity," Anderson said. "We had what amounted to a Hollywood movie set of a website. It barely got past the VCs."

Week by week, more engineers arrived and the vision became more real. The company secured a banking license and a mutual fund license and formed a partnership with Barclays. By November, X.com's small software team had created one of the world's first online banks complete with FDIC insurance to back the bank accounts and three mutual funds for investors to choose. Musk gave the engineers $100,000 of his own money to conduct their testing. On the night before Thanksgiving in 1999, X.com went live to the public. "I was there until two A.M.," Anderson said. "Then, I went home to cook Thanksgiving dinner. Elon called me a few hours later and asked me to come into the office to relieve some of the other engineers. Elon stayed there forty-eight straight hours, making sure things worked."

Under Musk's direction, X.com tried out some radical banking concepts. Customers received a $20 cash card just for signing up to use the service and a $10 card for every person they referred. Musk did away with niggling fees and overdraft penalties. In a very modern twist, X.com also built a person-to-person payment system in which you could send someone money just by plugging their e-mail address into the site. The whole idea was to shift away from slow-moving banks with their mainframes taking days to process payments and to create a kind of agile bank account where you could move money around with a couple of clicks on a mouse or an e-mail. This was revolutionary stuff, and more than 200,000 people bought into it and signed up for X.com within the first couple of months of operation.

Soon enough, X.com had a major competitor. A couple of brainy kids named Max Levchin and Peter Thiel had been working on a payment system of their own at their start-up called Confinity. The duo actually rented their office space—a glorified broom closet—from X.com and were trying to make it possible for owners of Palm Pilot handhelds to swap money via the infrared ports on the devices. Between X.com and Confinity, the small office on University Avenue had turned into the frenzied epicenter of the Internet finance revolution. "It was this mass of adolescent men that worked so hard," Ankenbrandt said. "It stunk so badly in there. I can still smell it—leftover pizza, body odor, and sweat."

The pleasantries between X.com and Confinity came to an abrupt end. The Confinity founders moved to an office down the street and, like X.com, began focusing their attention on Web- and e-mail-based payments with their service known as PayPal. The companies became locked in a heated battle to match each other's features and attract more users, knowing that whoever got bigger faster would win. Tens of millions of dollars were spent on promotions, while millions more were lost battling hackers who had seized upon the services as new playgrounds for fraud. "It was like the Internet version of making it rain at a strip club," said Jeremy Stoppelman, an X.com engineer who went on to become the CEO of Yelp. "You gave away money as fast as you could."

The race to win Internet payments gave Musk a chance to show off his quick thinking and work ethic. He kept devising plans to counter the advantage PayPal had established on auction sites like eBay. And he rallied the X.com employees to implement the tactics as fast as possible using brute-force appeals to their competitive natures. "There really wasn't anything suave about him," Ankenbrandt said. "We all worked twenty hours a day, and he worked twenty-three hours."

In March 2000, X.com and Confinity finally decided to stop trying to spend each other into oblivion and to join forces. Confinity had what looked like the hottest product in PayPal but was paying out $100,000 a day in awards to new customers and didn't have the cash reserves to keep going. X.com, by contrast, still had plenty of cash reserves and the more sophisticated banking products. It took the lead in setting the merger terms, leaving Musk as the largest shareholder of the combined company, which would be called X.com. Shortly after the deal closed, X.com raised $100 million from backers including Deutsche Bank and Goldman Sachs and boasted that it had more than one million customers.[*]

The two companies tried hard to mesh their cultures, with modest success. Groups of employees from X.com tied their computer monitors to their desk chairs with power cords and rolled them down the street to the Confinity offices to work alongside their new colleagues. But the teams could never quite see eye to eye. Musk kept championing the X.com brand, while most everyone else favored PayPal. More fights broke out over the design of the company's technology infrastructure. The Confinity team led by Levchin favored moving toward open-source software like Linux, while Musk championed Microsoft's data-center software as being more likely to keep productivity high. This squabble may sound silly to outsiders, but it was the equivalent of a religious war to the engineers, many of whom viewed Microsoft as a dated evil

---

[*] Musk had been pushed out as CEO of X.com by the company's investors, who wanted a more seasoned executive to lead the company toward an IPO. In December 1999, X.com hired Bill Harris, the former CEO of the financial software maker Intuit, as its new chief. After the merger, many in the company turned on Harris, he resigned, and Musk returned as the CEO.

empire and Linux as the modern software of the people. Two months after the merger, Thiel resigned and Levchin threatened to walk out over the technology rift. Musk was left to run a fractured company.

The technology issues X.com had been facing worsened as the computing systems failed to keep up with an exploding customer base. Once a week, the company's website collapsed. Most of the engineers were ordered to start work designing a new system, which distracted key technical personnel and left X.com vulnerable to fraud. "We were losing money hand over fist," said Stoppelman. As X.com became more popular and its transaction volume exploded, all of its problems worsened. There was more fraud. There were more fees from banks and credit card companies. There was more competition from start-ups. X.com lacked a cohesive business model to offset the losses and turn a profit from the money it managed. Roelof Botha, the start-up's chief financial officer and now a prominent venture capitalist at Sequoia, did not think Musk provided the board with a true picture of X.com's issues. A growing number of other people at the company questioned Musk's decision-making in the face of all the crises.

What followed was one of the nastiest coups in Silicon Valley's long, illustrious history of nasty coups. A small group of X.com employees gathered one night at Fanny & Alexander, a now-defunct bar in Palo Alto, and brainstormed about how to push out Musk. They decided to sell the board on the idea of Thiel returning as CEO. Instead of confronting Musk directly with this plan, the conspirators decided to take action behind Musk's back.

Musk and Justine had been married in January 2000 but had been too busy for a honeymoon. Nine months later, in September, they planned to mix business and pleasure by going on a fund-raising trip and ending it with a honeymoon in Sydney to catch the Olympics. As they boarded their flight one night, X.com executives delivered letters of no confidence to X.com's board. Some of the people loyal to Musk had sensed something was wrong, but it was too late. "I went to the office at ten thirty that night, and everyone was there," Ankenbrandt said. "I could not believe it. I am frantically trying to call Elon, but he's on a plane." By the time he landed, Musk had been replaced by Thiel.

When Musk finally heard what had happened, he hopped on the next plane back to Palo Alto. "It was shocking, but I will give Elon this—I thought he handled it pretty well," Justine said. For a brief period, Musk tried to fight back. He urged the board to reconsider its decision. But when it became clear that the company had already moved on, Musk relented. "I talked to Moritz and a few others," Musk said. "It wasn't so much that I wanted to be CEO but more like, 'Hey, I think there are some pretty important things that need to happen, and if I'm not CEO, I'm not sure they are going to happen.' But then I talked to Max and Peter, and it seemed like they would make these things happen. So then, I mean, it's not the end of the world."

Many of the X.com employees who had been with Musk since early on were less than impressed by what had happened. "I was floored by it and angry," said Stoppelman. "Elon was sort of a rock star in my view. I was very vocal about how I thought it was bullshit. But I knew fundamentally that the company was doing well. It was a rocket ship, and I wasn't going to leave." Stoppelman, then twenty-three, went into a conference room and tore into Thiel and Levchin. "They let me vent it all out, and their reaction was part of the reason I stayed." Others remained embittered. "It was backhanded and cowardly," said Branden Spikes, a Zip2 and X.com engineer. "I would have been more behind it if Elon had been in the room."

By June 2001, Musk's influence on the company was fading

quickly. That month, Thiel rebranded X.com as PayPal. Musk rarely lets a slight go unpunished. Throughout this ordeal, however, he showed incredible restraint. He embraced the role of being an advisor to the company and kept investing in it, increasing his stake as PayPal's largest shareholder. "You would expect someone in Elon's position to be bitter and vindictive, but he wasn't," said Botha. "He supported Peter. He was a prince."

The next few months would end up being key for Musk's future. The dot-com joyride was coming to a quick end, and people wanted to try to cash out in any way possible. When executives from eBay began approaching PayPal about an acquisition, the inclination for most people was to sell and sell fast. Musk and Moritz, though, urged the board to reject a number of offers and hold out for more money. PayPal had revenue of about $240 million per year, and looked like it might make it as an independent company and go public. Musk and Moritz's resistance paid off and then some. In July 2002, eBay offered $1.5 billion for PayPal, and Musk and the rest of the board accepted the deal. Musk netted about $250 million from the sale to eBay, or $180 million after taxes—enough to make what would turn out to be his very wild dreams possible.

The PayPal episode was a mixed bag for Musk. His reputation as a leader suffered in the aftermath of the deal, and the media turned on him in earnest for the first time. Eric Jackson, an early Confinity employee, wrote *The PayPal Wars: Battles with eBay, the Media, the Mafia, and the Rest of Planet Earth* in 2004 and recounted the company's tumultuous journey. The book painted Musk as an egomaniacal, stubborn jerk, making wrong decisions at every turn, and portrayed Thiel and Levchin as heroic geniuses. Valleywag, the technology industry gossip site, piled on as well and turned bashing Musk into one of its pet projects. The criticisms grew to the point that people started wondering aloud whether or not Musk counted as a true cofounder of PayPal or had just ridden Thiel's coattails to a magical payday. The tone of the book along with the blog posts goaded Musk in 2007 into writing a 2,200-word e-mail to Valleywag meant to set the record straight with his version of events.

In the e-mail, Musk let his literary flair loose and gave the public a direct look at his combative side. He described Jackson as "a sycophantic jackass" and "one notch above an intern," who had little insight into the high-level goings-on at the company. "Since Eric worships Peter, the outcome was obvious—Peter sounds like Mel Gibson in *Braveheart* and my role is somewhere between negligible and a bad seed," Musk wrote. Musk then detailed seven reasons why he deserved cofounder status of PayPal, including his role as its largest shareholder, the hiring of a lot of the top talent, the creation of a number of the company's most successful business ideas, and his time as CEO when the company went from sixty to several hundred employees.

Almost everyone I interviewed from the PayPal days leaned toward agreeing with Musk's overall assessment. They said that Jackson's account bordered on fantasy when it came to celebrating the Confinity team over Musk and the X.com team. "There are a lot of PayPal people that suffer from warped memories," said Botha.

But these same people reached another consensus, saying that Musk had mishandled the branding, technology infrastructure, and fraud situations. "I think it would have killed the company if Elon had stayed on as CEO for six more months," said Botha. "The mistakes Elon was making at the time were amplifying the risk of the business." (For more on Musk's take on the PayPal years, see Appendix 2.)

The suggestions that Musk did not count as a "true" cofounder of PayPal seem asinine in retrospect. Thiel, Levchin, and other

PayPal executives have said as much in the years since the eBay deal closed. The only useful thing such criticisms produced were the bombastic counteroffensives from Musk, which revealed touches of insecurity and the seriousness with which Musk insists that the historical record reflect his take on events. "He comes from the school of thought in the public relations world that you let no inaccuracy go uncorrected," said Vince Sollitto, the former communications chief at PayPal. "It sets a precedent, and you should fight every out-of-place comma tooth and nail. He takes things very personally and usually seeks war."

The stronger critique of Musk during this period of his life was that he had succeeded to a large degree despite himself. Musk's traits as a confrontational know-it-all and his abundant ego created deep, lasting fractures within his companies. While Musk consciously tried to temper his behavior, these efforts were not enough to win over investors and more experienced executives. At both Zip2 and PayPal, the companies' boards came to the conclusion that Musk was not yet CEO material. It can also be argued that Musk had become a hyperbolic huckster, who overreached and oversold his companies' technology. Musk's biggest detractors have made all of these arguments either in public or private and a half dozen or so of them said far worse things to me about his character and actions, describing Musk as unethical in business and vicious with his personal attacks. Almost universally, these people were unwilling to go on the record with their comments, claiming to be afraid Musk would pursue litigation against them or ruin their ability to do business.

These criticisms must be weighed against Musk's track record. He demonstrated an innate ability to read people and technology trends at the inception of the consumer Web. While others tried to wrap their heads around the Internet's implications, Musk had already set off on a purposeful plan of attack. He envisioned many of the early pieces of technology—directories, maps, sites that focused on vertical markets—that would become mainstays on the Web. Then, just as people became comfortable with buying things from Amazon.com and eBay, Musk made the great leap forward to full-fledged Internet banking. He would bring standard financial instruments online and then modernize the industry with a host of new concepts. He exhibited a deep insight into human nature that helped his companies pull off exceptional marketing, technology, and financial feats. Musk was already playing the entrepreneur game at the highest level and working the press and investors like few others could. Did he hype things up and rub people the wrong way? Absolutely—and with spectacular results.

Based in large part on Musk's guidance, PayPal survived the bursting of the dot-com bubble, became the first blockbuster IPO after the 9/11 attacks, and then sold to eBay for an astronomical sum while the rest of the technology industry was mired in a dramatic downturn. It was nearly impossible to survive let alone emerge as a winner in the midst of such a mess.

PayPal also came to represent one of the greatest assemblages of business and engineering talent in Silicon Valley history. Both Musk and Thiel had a keen eye for young, brilliant engineers. The founders of start-ups as varied as YouTube, Palantir Technologies, and Yelp all worked at PayPal. Another set of people—including Reid Hoffman, Thiel, and Botha—emerged as some of the technology industry's top investors. PayPal staff pioneered techniques in fighting online fraud that have formed the basis of software used by the CIA and FBI to track terrorists and of software used by the world's largest banks to combat crime. This collection of super-bright employees has become known as the PayPal Mafia—more or less the current ruling class of Silicon Valley—and Musk is its most famous and successful member.

Hindsight also continues to favor Musk's unbridled vision over the more cautious pragmatism of executives at Zip2 and PayPal. Had it chased consumers as Musk urged, Zip2 may have ended up as a blockbuster mapping and review service. As for PayPal, an argument can still be made that the investors sold out too early and should have listened more to Musk's demands to remain independent. By 2014, PayPal had amassed 153 million users and was valued at close to $32 billion as a stand-alone company. A flood of payment and banking start-ups have appeared as well—Square, Stripe, and Simple, to name three among the S's—that have looked to fulfill much of the original X.com vision.

If X.com's board had been a bit more patient with Musk, there's good reason to believe he would have succeeded with delivery of the "online bank to rule them all" that he had set out to create. History has demonstrated that while Musk's goals can sound absurd in the moment, he certainly believes in them and, when given enough time, tends to achieve them. "He always works from a different understanding of reality than the rest of us," Ankenbrandt said. "He is just different than the rest of us."

While navigating the business tumult of Zip2 and PayPal, Musk found a moment of peace in his personal life. He'd spent years courting Justine Wilson from afar, flying her out for visits on the weekends. For a long time, his oppressive hours and his roommates put a crimp on the relationship. But the Zip2 sale let Musk buy a place of his own and pay a bit more attention to Justine. Like any couple, they had their ups and downs, but that passion of young love remained. "We fought a lot, but when we weren't fighting, there was a deep sense of compassion—a bond," Justine said. The couple had been sparring for a few days about phone calls Justine kept getting from an ex-boyfriend—"Elon didn't like that"—and had a major spat while walking near the X.com offices. "I remember thinking it was a lot of drama, and that if I was going to put up with it, we might as well be married. I told him he should just propose to me," Justine said. It took Musk a few minutes to cool down and then he did just that, proposing on the spot. A few days later, a more chivalrous Musk returned to the sidewalk, got down on bended knee, and presented Justine with a ring.

Justine knew all about Musk's grim childhood and the intense range of emotions he could exhibit. Her romantic sensibilities overrode any trepidation she might have had about these parts of Musk's history and character and centered instead on his strength. Musk often talked fondly about Alexander the Great, and Justine saw him as her own conquering hero. "He wasn't afraid of responsibility," she said. "He didn't run from things. He wanted to get married and have kids early on." Musk also exuded a confidence and passion that made Justine think life with him would always be okay. "Money is not his motivation, and, quite frankly, I think it just happens for him," Justine said. "It's just there. He knows he can generate it."

At their wedding reception, Justine encountered the other side of the conquering hero. Musk pulled Justine close while they danced, and informed her, "I am the alpha in this relationship."[3] Two months later, Justine signed a postnuptial financial agreement that would come back to haunt her and entered into an enduring power struggle. She described the situation years later in an article for *Marie Claire*, writing, "He was constantly remarking on the ways he found me lacking. 'I am your wife,' I told him repeatedly, 'not your employee.' 'If you were my employee,' he said just as often, 'I would fire you.'"

The newlyweds were not helped by the drama at X.com. They'd put off their honeymoon and then had it derailed by the coup. It took until late December 2000 for things to calm down enough for Musk to take his first vacation in years. He arranged a two-week trip, with the first part taking place in Brazil and the second in South Africa at a game reserve near the Mozambique border. While in Africa, Musk contracted the most virulent version of malaria—falciparum malaria—which accounts for the vast majority of malaria deaths.

Musk returned to California in January, which is when the illness took hold. He started to get sick and was bedridden for a few days before Justine took him to a doctor who then ordered that Musk be rushed in an ambulance to Sequoia Hospital in Redwood City.* Doctors there misdiagnosed and mistreated his condition to the point that Musk was near death. "Then, there happened to be a guy visiting from another hospital who had seen a lot more malaria cases," Musk said. He spied Musk's blood work in the lab and ordered an immediate maximum dosage of doxycycline, an antibiotic. The doctor told Musk that if he had turned up a day later, the medicine likely would no longer have been effective.

---

* After feeling ill for a few days, Musk went to Stanford Hospital and informed them that he'd been in a malaria zone, although the doctors could not find the parasite during tests. The doctors performed a spinal tap and diagnosed him with viral meningitis. "I may very well have also had that, and they treated me for it, and it did get better," Musk said. The doctors discharged Musk from the hospital and warned him that some symptoms would recur. "I started feeling bad a few days later, and it got progressively worse," Musk said. "Eventually, I couldn't walk. It was like, 'Okay, this is even worse than the first time.'" Justine took Musk to a general practitioner in a cab, and he lay on the floor of the doctor's office. "I was so dehydrated that she couldn't take my vitals," Musk said. The doctor called an ambulance, which transported Musk to Sequoia Hospital in Redwood City with IVs in both arms. Musk faced another misdiagnosis—this time of the type of malaria. The doctors declined to give Musk a more aggressive treatment that came with nasty side effects including heart palpitations and organ failure.

Musk spent ten agonizing days in the intensive care unit. The experience shocked Justine. "He's built like a tank," she said. "He has a level of stamina and an ability to deal with levels of stress that I've never seen in anyone else. To see him laid low like that in total misery was like a visit to an alternate universe." It took Musk six months to recover. He lost forty-five pounds over the course of the illness and had a closet full of clothes that no longer fit. "I came very close to dying," Musk said. "That's my lesson for taking a vacation: vacations will kill you."

6

MICE IN SPACE

ELON MUSK TURNED THIRTY IN JUNE 2001, AND THE BIRTHDAY HIT HIM hard. "I'm no longer a child prodigy," he told Justine, only half joking. That same month X.com officially changed its name to PayPal, providing a harsh reminder that the company had been ripped away from Musk and given to someone else to run. The start-up life, which Musk described as akin to "eating glass and staring into the abyss,"⁴ had gotten old and so had Silicon Valley. It felt like Musk was living inside a trade show where everyone worked in the technology industry and talked all the time about funding, IPOs, and chasing big paydays. People liked to brag about the crazy hours they worked, and Justine would just laugh, knowing Musk had lived a more extreme version of the Silicon Valley lifestyle than they could imagine. "I had friends who complained that their husbands came home at seven or eight," she said. "Elon would come home at eleven and work some more. People didn't always get the sacrifice he made in order to be where he was."

The idea of escaping this incredibly lucrative rat race started to grow more and more appealing. Musk's entire life had been about chasing a bigger stage, and Palo Alto seemed more like a stepping-stone than a final destination. The couple decided to move south and begin their family and the next chapter of their lives in Los Angeles.

"There's an element to him that likes the style and the excitement and color of a place like L.A.," said Justine. "Elon likes to be where the action is." A small group of Musk's friends who felt similarly had also decamped to Los Angeles for what would be a wild couple of years.

It wasn't just Los Angeles's glitz and grandeur that attracted Musk. It was also the call of space. After being pushed out of PayPal, Musk had started to revisit his childhood fantasies around rocket ships and space travel and to think that he might have a greater calling than creating Internet services. The changes in his attitude and thinking soon became obvious to his friends, including a group of PayPal executives who had gathered in Las Vegas one weekend to celebrate the company's success. "We're all hanging out in this cabana at the Hard Rock Cafe, and Elon is there reading some obscure Soviet rocket manual that was all moldy and looked like it had been bought on eBay," said Kevin Hartz, an early PayPal investor. "He was studying it and talking openly about space travel and changing the world."

Musk had picked Los Angeles with intent. It gave him access to space or at least the space industry. Southern California's mild, consistent weather had made it a favored city of the aeronautics industry since the 1920s, when the Lockheed Aircraft Company set up shop in Hollywood. Howard Hughes, the U.S. Air Force, NASA, Boeing, and myriad other people and organizations have performed much of their manufacturing and cutting-edge experimentation in and around Los Angeles. Today the city remains a major hub for the military's aeronautics work and commercial activity. While Musk didn't know exactly what he wanted to do in space, he realized that just by being in Los Angeles he would be surrounded by the world's top aeronautics thinkers. They could help him refine any ideas, and there would be plenty of recruits to join his next venture.

Musk's first interactions with the aeronautics community were with an eclectic collection of space enthusiasts, members of a nonprofit group called the Mars Society. Dedicated to exploring and settling the Red Planet, the Mars Society planned to hold a fund-raiser in mid-2001. The $500-per-plate event was to take place at the house of one of the well-off Mars Society members, and invitations to the usual characters had been mailed out. What stunned Robert Zubrin, the head of the group, was the reply from someone named Elon Musk, whom no one could remember inviting. "He gave us a check for five thousand dollars," Zubrin said. "That made everyone take notice." Zubrin began researching Musk, determined he was rich, and invited him for coffee ahead of the dinner. "I wanted to make sure he knew the projects we had under way," Zubrin said. He proceeded to regale Musk with tales of the research center the society had built in the Arctic to mimic the tough conditions of Mars and the experiments they had been running for something called the Translife Mission, in which there would be a spinning capsule orbiting Earth that was piloted by a crew of mice. "It would spin to give them one-third gravity—the same you would have on Mars—and they would live there and reproduce," Zubrin told Musk.

When it was time for dinner, Zubrin placed Musk at the VIP table next to himself, the director and space buff James Cameron, and Carol Stoker, a planetary scientist for NASA with a deep interest in Mars. "Elon is so youthful-looking and at that time he looked like a little boy," Stoker said. "Cameron was chatting him up right away to invest in his next movie, and Zubrin was trying to get him to make a big donation to the Mars Society." In return for being hounded for cash, Musk probed about for ideas and contacts. Stoker's husband was an aerospace engineer at NASA working on a concept for an airplane that would glide over Mars looking for water. Musk loved that. "He was much more intense than some of the other millionaires," Zubrin said. "He didn't know a lot about space, but he had a scientific mind. He wanted to know exactly what was being planned in regards to Mars and what the significance would be." Musk took to the Mars Society right away and joined its board of directors. He donated another $100,000 to fund a research station in the desert as well.

Musk's friends were not entirely sure what to make of his mental state. He'd lost a tremendous amount of weight fighting off malaria and looked almost skeletal. With little prompting, Musk would start expounding on his desire to do something meaningful with his life—something lasting. His next move had to be either in solar or in space. "He said, 'The logical thing to happen next is solar, but I can't figure out how to make any money out of it,'" said George Zachary, the investor and close friend of Musk's, recalling a lunch date at the time. "Then he started talking about space, and I thought he meant office space like a real estate play." Musk had actually started thinking bigger than the Mars Society. Rather than send a few mice into Earth's orbit, Musk wanted to send them to Mars. Some very rough calculations done at the time suggested that the journey would cost $15 million. "He asked if I thought that was crazy," Zachary said. "I asked, 'Do the mice come back? Because, if they don't, yeah, most people will think that's crazy.'" As it turned out, the mice were not only meant to go to Mars and come back but were also meant to procreate along the way, during a journey that would take months. Jeff Skoll, another one of Musk's friends who made a fortune at eBay, pointed out that the fornicating mice would need a hell of a lot of cheese and bought Musk a giant wheel of Le Brouère, a type of Gruyère.

Musk did not mind becoming the butt of cheese jokes. The more he thought about space, the more important its exploration seemed to him. He felt as if the public had lost some of its ambition and hope for the future. The average person might see space exploration as a waste of time and effort and rib him for talking about the subject, but Musk thought about interplanetary travel in a very earnest way. He wanted to inspire the masses and reinvigorate their passion for science, conquest, and the promise of technology.

His fears that mankind had lost much of its will to push the boundaries were reinforced one day when Musk went to the NASA website. He'd expected to find a detailed plan for exploring Mars and instead found bupkis. "At first I thought, jeez, maybe I'm just looking in the wrong place," Musk once told *Wired*. "Why was there no plan, no schedule? There was nothing. It seemed crazy." Musk believed that the very idea of America was intertwined with humanity's desire to explore. He found it sad that the American agency tasked with doing audacious things in space and exploring new frontiers as its mission seemed to have no serious interest in investigating Mars at all. The spirit of Manifest Destiny had been deflated or maybe even come to a depressing end, and hardly anyone seemed to care.

Like so many quests to revitalize America's soul and bring hope to all of mankind, Musk's journey began in a hotel conference room. By this time, Musk had built up a decent network of contacts in the space industry, and he brought the best of them together at a series of salons—sometimes at the Renaissance hotel at the Los Angeles airport and sometimes at the Sheraton hotel in Palo Alto. Musk had no formal business plan for these people to debate. He mostly wanted them to help him develop the mice-to-Mars idea or at least to come up with something comparable. Musk hoped to hit on a grand gesture for mankind—some type of event that would capture the world's attention, get people thinking about Mars again, and have them reflect on man's potential. The scientists and luminaries at the meetings were to figure out a spectacle that would be technically feasible at a price tag of approximately $20 million. Musk resigned from his position as a director of the Mars Society and announced his own organization—the Life to Mars Foundation.

The collection of talent attending these sessions in 2001 was impressive. Scientists showed up from NASA's Jet Propulsion Laboratory, or JPL. James Cameron appeared, lending some celebrity to the affair. Also attending was Michael Griffin, whose academic credentials were spectacular and included degrees in aerospace engineering, electrical engineering, civil engineering, and applied physics. Griffin had worked for the CIA's venture capital arm called In-Q-Tel, at NASA, and at JPL and was just in the process of leaving Orbital Sciences Corporation, a maker of satellites and spacecraft, where he had been chief technical officer and the general manager of the space systems group. It could be argued that no one on the planet knew more about the realities of getting things into space than Griffin, and he was working for Musk as space thinker in chief. (Four years later, in 2005, Griffin took over as head of NASA.)

The experts were thrilled to have another rich guy appear who was willing to fund something interesting in space. They happily debated the merits and feasibility of sending up rodents and watching them hump. But, as the discussion wore on, a consensus started to build around pursuing a different project—something called "Mars Oasis." Under this plan, Musk would buy a rocket and use it to shoot what amounted to a robotic greenhouse to Mars. A group of researchers had already been working on a space-ready growth chamber for plants. The idea was to modify their structure, so that it could open up briefly and suck in some of the Martian regolith, or soil, and then use it to grow a plant, which would in turn produce the first oxygen on Mars. Much to Musk's liking, this new plan seemed both ostentatious and feasible.

Musk wanted the structure to have a window and a way to send a video feedback to Earth, so that people could watch the plant grow. The group also talked about sending out kits to students around the country who would grow their own plants simultaneously and take notice, for example, that the Martian plant could grow twice as high as its Earth-bound counterpart in the same amount of time. "This concept had been floating around in various forms for a while," said Dave Bearden, a space industry veteran who attended the meetings. "It would be, yes, there is life on Mars, and we put it there. The hope was that it might turn on a light for thousands of kids that this place is not that hostile. Then they might start thinking, Maybe we should go there." Musk's enthusiasm for the idea started to inspire the group, many of whom had grown cynical about anything novel happening in space again. "He's a very smart, very driven guy with a huge ego," Bearden said. "At one point someone mentioned that he might become *Time* magazine's Man of the Year, and you could see him light up. He has this belief that he is the guy who can change the world."

The main thing troubling the space experts was Musk's budget. Following the salons, it seemed like Musk wanted to spend somewhere between $20 million and $30 million on the stunt, and everyone knew that the cost of a rocket launch alone would eat up that money and then some. "In my mind, you needed two hundred million dollars to do it right," Bearden said. "But people were reluctant to bring too much reality into the situation too early and just get the whole idea killed." Then there were the immense engineering challenges that would need solving. "To have a big window on this thing was a real thermal problem," Bearden said. "You could not keep the container warm enough to keep anything alive." Scooping Martian soil into the structure seemed not only hard to do physically but also like a flat-out bad idea since the regolith would be toxic. For a while, the scientists debated growing the plant in a nutrient-rich gel instead, but that felt like cheating and like it might undermine the whole point of the endeavor. Even the optimistic moments were awash in unknowns. One scientist found some very resilient mustard seeds and thought they could possibly survive a treated version of the Martian soil. "There was a pretty big downside if the plant didn't survive," Bearden said. "You have this dead garden on Mars that ends up giving off the opposite of the intended effect."*

Musk never flinched. He turned some of the volunteer thinkers into consultants, and put them to work on the plant machine's design. He also plotted a trip to Russia to find out exactly how much a launch would cost. Musk intended to buy a refurbished intercontinental ballistic missile, or ICBM, from the Russians and use that as his launch vehicle. For help with this, Musk reached out to Jim Cantrell, an unusual fellow who had done a mix of classified and unclassified work for the United States and other governments. Among other claims to fame, Cantrell had been accused of espionage and placed under house arrest in 1996 by the Russians after a satellite deal went awry. "After a couple of

---

*   When Zubrin and some of the other Mars buffs heard of Musk's plant project, they were upset. "It didn't make any sense," Zubrin said. "It was a purely symbolic thing to do, and the second they opened that door, millions of microbes would escape and plague all of NASA's contamination protocols."

weeks, Al Gore made some calls, and it got worked out," Cantrell said. "I didn't want anything to do with the Russians again—ever." Musk had other ideas.

Cantrell was driving his convertible on a hot July evening in Utah when a call came in. "This guy in a funny accent said, 'I really need to talk to you. I am a billionaire. I am going to start a space program.'" Cantrell could not hear Musk well—he thought his name was Ian Musk—and said he would call back once he got home. The two men didn't exactly trust each other at the outset. Musk refused to give Cantrell his cell phone number and made the call from his fax machine. Cantrell found Musk both intriguing and all too eager. "He asked if there was an airport near me and if I could meet the next day," Cantrell said. "My red flags started going off." Fearing one of his enemies was trying to orchestrate an elaborate setup, Cantrell told Musk to meet him at the Salt Lake City airport, where he would rent a conference room near the Delta lounge. "I wanted him to meet me behind security so he couldn't pack a gun," Cantrell said. When the meeting finally took place, Musk and Cantrell hit it off. Musk rolled out his "humans need to become a multiplanetary species" speech, and Cantrell said that if Musk was really serious, he'd be willing to go to Russia—again—and help buy a rocket.

In late October 2001, Musk, Cantrell, and Adeo Ressi, Musk's friend from college, boarded a commercial flight to Moscow. Ressi had been playing the role of Musk's guardian and trying to ascertain whether his best friend had started to lose his mind. Compilation videos of rockets exploding were made, and interventions were held with Musk's friends trying to talk him out of wasting his money. While these measures failed, Adeo went along to Russia to try to contain Musk as best as he could. "Adeo would call me to the side and say, 'What Elon is doing is insane. A philanthropic gesture? That's crazy,'" Cantrell said. "He was seriously worried but was down with the trip." And why not? The men were heading to Russia at the height of its freewheeling post-Soviet days when rich guys could apparently buy space missiles on the open market.

Team Musk would grow to include Mike Griffin, and meet with the Russians three times over a period of four months.* The group set up a few meetings with companies like NPO Lavoch-kin, which had made probes intended for Mars and Venus for the Russian Federal Space Agency, and Kosmotras, a commercial rocket launcher. The appointments all seemed to go the same way, following Russian decorum. The Russians, who often skip breakfast, would ask to meet around 11 A.M. at their offices for an early lunch. Then there would be small talk for an hour or more as the meeting attendees picked over a spread of sandwiches, sausages, and, of course, vodka. At some point during this process, Griffin usually started to lose his patience. "He suffers fools very poorly," Cantrell said. "He's looking around and wondering when we're going to get down to fucking business." The answer was not soon. After lunch came a lengthy smoking and coffee-drinking period. Once all of the tables were cleared, the Russian in charge would turn to Musk and ask, "What is it you're inter-ested in buying?" The big windup may not have bothered Musk as much if the Russians had taken him more seriously. "They looked at us like we were not credible people," Cantrell said. "One of their chief designers spit on me and Elon because he thought we were full of shit."

---

* Most of the stories written about Musk that touch on this period say he went to Moscow three times. According to Cantrell's detailed records, this is not the case. Musk met with the Russians twice in Moscow, and once in Pasadena, California. He also met with Arianespace in Paris, and in London with Surrey Satellite Technology Ltd., which Musk considered buying.

The most intense meeting occurred in an ornate, neglected, prerevolutionary building near downtown Moscow. The vodka shots started—"To space!" "To America!"—while Musk sat on $20 million, which he hoped would be enough to buy three ICBMs that could be retooled to go to space. Buzzed from the vodka, Musk asked point-blank how much a missile would cost. The reply: $8 million each. Musk countered, offering $8 million for two. "They sat there and looked at him," Cantrell said. "And said something like, 'Young boy. No.' They also intimated that he didn't have the money." At this point, Musk had decided the Russians were either not serious about doing business or determined to part a dot-com millionaire from as much of his money as possible. He stormed out of the meeting.

The Team Musk mood could not have been worse. It was near the end of February 2002, and they went outside to hail a cab and drove straight to the airport surrounded by the snow and dreck of the Moscow winter. Inside the cab, no one talked. Musk had come to Russia filled with optimism about putting on a great show for mankind and was now leaving exasperated and disappointed by human nature. The Russians were the only ones with rockets that could possibly fit within Musk's budget. "It was a long drive," Cantrell said. "We sat there in silence looking at the Russian peasants shopping in the snow." The somber mood lingered all the way to the plane, until the drink cart arrived. "You always feel particularly good when the wheels lift off in Moscow," Cantrell said. "It's like, 'My God. I made it.' So, Griffin and I got drinks and clinked our glasses." Musk sat in the row in front of them, typing on his computer. "We're thinking, Fucking nerd. What can he be doing now?" At which point Musk wheeled around and flashed a spreadsheet he'd created. "Hey, guys," he said, "I think we can build this rocket ourselves."

Griffin and Cantrell had downed a couple of drinks by this time and were too deflated to entertain a fantasy. They knew all too well the stories of gung-ho millionaires who thought they could conquer space only to lose their fortunes. Just the year before, Andrew Beal, a real estate and finance whiz in Texas, folded his aerospace company after having poured millions into a massive test site. "We're thinking, Yeah, you and whose fuck-ing army," Cantrell said. "But, Elon says, 'No, I'm serious. I have this spreadsheet.'" Musk passed his laptop over to Griffin and Cantrell, and they were dumbfounded. The document detailed the costs of the materials needed to build, assemble, and launch a rocket. According to Musk's calculations, he could undercut existing launch companies by building a modest-sized rocket that would cater to a part of the market that specialized in carrying smaller satellites and research payloads to space. The spreadsheet also laid out the hypothetical performance characteristics of the rocket in fairly impressive detail. "I said, 'Elon, where did you get this?'" Cantrell said.

Musk had spent months studying the aerospace industry and the physics behind it. From Cantrell and others, he'd borrowed *Rocket Propulsion Elements*, *Fundamentals of Astrodynamics*, and *Aerothermodynamics of Gas Turbine and Rocket Propulsion*, along with several more seminal texts. Musk had reverted to his child-hood state as a devourer of information and had emerged from this meditative process with the realization that rockets could and should be made much cheaper than what the Russians were offering. Forget the mice. Forget the plant with its own video feed growing—or possibly dying—on Mars. Musk would inspire people to think about exploring space again by making it cheaper to explore space.

As word traveled around the space community about Musk's plans, there was a collective ho-hum. People like Zubrin had seen this show many times before. "There was a string of zillionaires that got sold a good story by an engineer," Zubrin said. "Combine my brains and your money, and we can build a rocket ship that will be profitable and open up the space frontier. The techies usually ended up spending the rich guy's money for two years, and then the rich guy gets bored and shuts the thing down. With Elon, everyone gave a sigh and said, 'Oh well. He could have spent ten million dollars to send up the mice, but instead he'll spend hundreds of millions and probably fail like all the others that proceeded him.'"

While well aware of the risks tied to starting a rocket company, Musk had at least one reason to think he might succeed where others had failed. That reason's name was Tom Mueller.

Mueller grew up the son of a logger in the tidy Idaho town of St. Maries, where he developed a reputation as an oddball. While the rest of the kids were outside exploring the woods in winter, Mueller stayed warm in the library reading books or watching *Star Trek* at his house. He also tinkered. Walking to grade school one day, Mueller discovered a smashed clock in an alley and turned it into a pet project. Each day, he fixed some part of the clock—a gear, a spring—until he got it working. A similar thing happened with the family's lawn mower, which Mueller disassembled one afternoon on the front lawn for fun. "My dad came home and was so mad because he thought he'd have to buy a new mower," Mueller said. "But I put it back together, and it ran." Mueller then got stuck on rockets. He started buying mail order kits and following the instructions to build small rockets. Rather quickly, Mueller graduated to constructing his own devices. At the age of twelve, he crafted a mock-up space shuttle that could be attached to a rocket, sent up into the air, and then glide back to the ground. For a science project a couple of years later, Mueller borrowed his dad's oxyacetylene welding equipment to make a rocket engine prototype. Mueller cooled the device by placing it upside down in a coffee can full of water—"I could run it like that all day long"—and invented equally creative ways to measure its performance. The machine was good enough for Mueller to win a couple of regional science fair competitions and end up at an international event. "That's where I promptly got my ass kicked," Mueller said.

Tall, lanky, and with a rectangular face, Mueller is an easy-going sort who muddled through college for a bit, teaching his friends how to make smoke bombs, and then eventually settled down and did well as a mechanical engineering student. Fresh out of college, he worked for Hughes Aircraft on satellites—"It wasn't rockets, but it was close"—and then went to TRW Space & Electronics. It was the latter half of the 1980s, and Ronald Reagan's Star Wars program had the space gearheads dreaming about kinetic weapons and all sorts of mayhem. At TRW, Mueller experimented with crazy types of propellants and oversaw the development of the company's TR-106 engine, a giant machine fueled by liquid oxygen and hydrogen. As a hobby, Mueller hung out with a couple hundred amateur rocketry buffs in the Reaction Research Society, a group formed in 1943 to encourage the building and firing of rockets. On the weekends, Mueller traveled out to the Mojave Desert with the other RRS members to push the limits of amateur machines. Mueller was one of the club's standouts, able to build things that actually worked, and could experiment with some of the more radical concepts that were quashed by his conservative bosses at TRW. His crowning achievement was an eighty-pound engine that could produce thirteen thousand pounds of thrust and earned accolades as the world's largest liquid-fuel rocket engine built by an amateur. "I still keep the rockets hanging in my garage," Mueller said.

In January 2002, Mueller was hanging out in the workshop

of John Garvey, who had left a job at the aerospace company McDonnell Douglas to start building his own rockets. Garvey's facility was in Huntington Beach, where he rented an industrial space about the size of a six-car garage. The two men were fiddling around with the eighty-pound engine when Garvey mentioned that a guy named Elon Musk might be stopping by. The amateur rocketry scene is tight, and it was Cantrell who recommended that Musk check out Garvey's workshop and see Mueller's designs. On a Sunday, Musk arrived with a pregnant Justine, wearing a stylish black leather trench coat and looking like a high-paid assassin. Mueller had the eighty-pound engine on his shoulder and was trying to bolt it to a support structure when Musk began peppering him with questions. "He asked me how much thrust it had," Mueller said. "He wanted to know if I had ever worked on anything bigger. I told him that yeah, I'd worked on a 650,000-pound thrust engine at TRW and knew every part of it." Mueller set the engine down and tried to keep up with Musk's interrogation. "How much would that big engine cost?" Musk asked. Mueller told him TRW built it for about $12 million. Musk shot back, "Yeah, but how much could you really do it for?"

Mueller ended up chatting with Musk for hours. The next weekend, Mueller invited Musk to his house to continue their discussion. Musk knew he had found someone who really knew the ins and outs of making rockets. After that, Musk introduced Mueller to the rest of his roundtable of space experts and their stealthy meetings. The caliber of the people impressed Mueller, who had turned down past job offers from Beal and other budding space magnates because of their borderline insane ideas. Musk, by contrast, seemed to know what he was doing, weeding out the naysayers meeting by meeting and forming a crew of bright, committed engineers.

Mueller had helped Musk fill out that spreadsheet around the performance and cost metrics of a new, low-cost rocket, and, along with the rest of Team Musk, had subsequently refined the idea. The rocket would not carry truck-sized satellites like some of the monster rockets flown by Boeing, Lockheed, the Russians, and others countries. Instead, Musk's rocket would be aimed at the lower end of the satellite market, and it could end up as ideal for an emerging class of smaller payloads that capitalized on the massive advances that had taken place in recent years in computing and electronics technology. The rocket would cater directly to a theory in the space industry that a whole new market might open for both commercial and research payloads if a company could drastically lower the price per launch and perform launches on a regular schedule. Musk relished the idea of being at the forefront of this trend and developing the workhorse of a new era in space. Of course, all of this was theoretical—and then, suddenly, it wasn't. PayPal had gone public in February with its shares shooting up 55 percent, and Musk knew that eBay wanted to buy the company as well. While noodling on the rocket idea, Musk's net worth had increased from tens of millions to hundreds of millions. In April 2002, Musk fully abandoned the publicity-stunt idea and committed to building a commercial space venture. He pulled aside Cantrell, Griffin, Mueller, and Chris Thompson, an aerospace engineer at Boeing, and told the group, "I want to do this company. If you guys are in, let's do it." (Griffin wanted to join but ended up declining when Musk rebuffed his request to live on the East Coast, and Cantrell only stuck around for a few months after this meeting, seeing the venture as too risky.)

Founded in June 2002, Space Exploration Technologies came to life in humble settings. Musk acquired an old warehouse at 1310 East Grand Avenue in El Segundo, a suburb of Los Angeles humming with the activity of the aerospace industry. The previous tenant of the 75,000-square-foot building had done lots of shipping and had used the south side of the facility as a logistics depot, outfitting it with several receiving bays for delivery trucks. This allowed Musk to drive his silver McLaren right into the building. Beyond that the surroundings were sparse—just a dusty floor and a forty-foot-high ceiling with its wooden beams and insulation exposed and which curved at the top to give the place a hangarlike feel. The north side of the building was an office space with cubicles and room for about fifty people. During the first week of SpaceX's operations, delivery trucks showed up packed full of Dell laptops and printers and folding tables that would serve as the first desks. Musk walked over to one of the loading docks, rolled up the door, and off-loaded the equipment himself.

Musk had soon transformed the SpaceX office with what has become his signature factory aesthetic: a glossy epoxy coating applied over concrete on the floors, and a fresh coat of white paint slathered onto the walls. The white color scheme was intended to make the factory look clean and feel cheerful. Desks were interspersed around the factory so that Ivy League computer scientists and engineers designing the machines could sit with the welders and machinists building the hardware. This approach stood as SpaceX's first major break with traditional aerospace companies that prefer to cordon different engineering groups off from each other and typically separate engineers and machinists by thousands of miles by placing their factories in locations where real estate and labor run cheap.

As the first dozen or so employees came to the offices, they were told that SpaceX's mission would be to emerge as the "Southwest Airlines of Space." SpaceX would build its own engines and then contract with suppliers for the other components of the rocket. The company would gain an edge over the competition by building a better, cheaper engine and by fine-tuning the assembly process to make rockets faster and cheaper than anyone else. This vision included the construction of a type of mobile launch vehicle that could travel to various sites, take the rocket from a horizontal to vertical position, and send it off to space—no muss, no fuss. SpaceX was meant to get so good at this process that it could do multiple launches a month, make money off each one, and never need to become a huge contractor dependent on government funds.

SpaceX was to be America's attempt at a clean slate in the rocket business, a modernized reset. Musk felt that the space industry had not really evolved in about fifty years. The aerospace companies had little competition and tended to make supremely expensive products that achieved maximum performance. They were building a Ferrari for every launch, when it was possible that a Honda Accord might do the trick. Musk, by contrast, would apply some of the start-up techniques he'd learned in Silicon Valley to run SpaceX lean and fast and capitalize on the huge advances in computing power and materials that had taken place over the past couple of decades. As a private company, SpaceX would also avoid the waste and cost overruns associated with government contractors. Musk declared that SpaceX's first rocket would be called the Falcon 1, a nod to *Star Wars*' Millennium Falcon and his role as the architect of an exciting future. At a time when the cost of sending a 550-pound payload started at $30 million, he promised that the Falcon 1 would be able to carry a 1,400-pound payload for $6.9 million.

Bowing to his nature, Musk set an insanely ambitious timeline for all of this. One of the earliest SpaceX presentations suggested that the company would complete its first engine in May 2003, a second engine in June, the body of the rocket in July, and have everything assembled in August. A launchpad would then be prepared by September, and the first launch would take place in November 2003, or about fifteen months after the company started. A trip to Mars was naturally slated for somewhere near the end of the decade. This was Musk the logical, naïve optimist tabulating how long it should take people physically to perform all of this work. It's the baseline he expects of himself and one that his employees, with their human foibles, are in a never-ending struggle to match.

As space enthusiasts started to learn about the new company, they didn't really obsess over whether Musk's delivery schedule sounded realistic or not. They were just thrilled that someone had decided to take the cheap and fast approach. Some members of the military had already been promoting the idea of giving the armed forces more aggressive space capabilities, or what they called "responsive space." If a conflict broke out, the military wanted the ability to respond with purpose-built satellites for that mission. This would mean moving away from a model where it takes ten years to build and deploy a satellite for a specific job. Instead, the military desired cheaper, smaller satellites that could be reconfigured through software and sent up on short notice, almost like disposable satellites. "If we could pull that off, it would be really game-changing," said Pete Worden, a retired air force general, who met with Musk while serving as a consultant to the Defense Department. "It could make our response in space similar to what we do on land, sea and in the air." Worden's job required him to look at radical technologies. While many of the people he encountered came off as eccentric dreamers, Musk seemed grounded, knowledgeable, and capable. "I talked to people building ray guns and things in their garages. It was clear that Elon was different. He was a visionary who really understood the rocket technology, and I was impressed with him."

Like the military, scientists wanted cheap, quick access to space and the ability to send up experiments and get data back on a regular basis. Some companies in the medical and consumer-goods industries were also interested in rides to space to study how a lack of gravity affected the properties of their products.

As good as a cheap launch vehicle sounded, the odds of a private citizen building one that worked were beyond remote. A quick search on YouTube for "rocket explosions" turns up thousands of compilation videos documenting U.S. and Soviet launch disasters that have occurred over the decades. From 1957 to 1966, the United States alone tried to blast more than 400 rockets into orbit and about 100 of them crashed and burned.[5] The rockets used to transport things to space are mostly modified missiles developed through all of this trial and error and funded by billions upon billions of government dollars. SpaceX had the advantage of being able to learn from this past work and having a few people on staff that had overseen rocket projects at companies like Boeing and TRW. That said, the start-up did not have a budget that could support a string of explosions. At best, SpaceX would have three or four shots at making the Falcon 1 work. "People thought we were just crazy," Mueller said. "At TRW, I had an army of people and government funding. Now we were going to make a low-cost rocket from scratch with a small team. People just didn't think it could be done."

In July 2002, Musk was gripped by the excitement of this daring enterprise, and eBay made its aggressive move to buy PayPal for $1.5 billion. This deal gave Musk some liquidity and supplied him with more than $100 million to throw at SpaceX. With such a massive up-front investment, no one would be able to wrestle control of SpaceX away from Musk as they had done at Zip2 and PayPal. For the employees who had agreed to accompany Musk on this seemingly impossible journey, the windfall provided at least a couple of years of job security. The acquisition also upped Musk's profile and celebrity, which he could leverage to score meetings with top government officials and to sway suppliers.

And then all of a sudden none of this seemed to matter. Justine had given birth to a son—Nevada Alexander Musk. He was ten weeks old when, just as the eBay deal was announced, he died. The Musks had tucked Nevada in for a nap and placed the boy on his back as parents are taught to do. When they returned to check on him, he was no longer breathing and had suffered from what the doctors would term a sudden infant death syndrome–related incident. "By the time the paramedics resuscitated him, he had been deprived of oxygen for so long that he was brain-dead," Justine wrote in her article for *Marie Claire*. "He spent three days on life support in a hospital in Orange County before we made the decision to take him off it. I held him in my arms when he died. Elon made it clear that he did not want to talk about Nevada's death. I didn't understand this, just as he didn't understand why I grieved openly, which he regarded as 'emotionally manipulative.' I buried my feelings instead, coping with Nevada's death by making my first visit to an IVF clinic less than two months later. Elon and I planned to get pregnant again as swiftly as possible. Within the next five years, I gave birth to twins, then triplets." Later, Justine chalked up Musk's reaction to a defense mechanism that he'd learned from years of suffering as a kid. "He doesn't do well in dark places," she told *Esquire* magazine. "He's forward-moving, and I think it's a survival thing with him."

Musk did open up to a couple of close friends and expressed the depth of his misery. But for the most part, Justine read her husband right. He didn't see the value in grieving publicly. "It made me extremely sad to talk about it," Musk said. "I'm not sure why I'd want to talk about extremely sad events. It does no good for the future. If you've got other kids and obligations, then wallowing in sadness does no good for anyone around you. I'm not sure what should be done in such situations."

Following Nevada's death, Musk threw himself at SpaceX and rapidly expanded the company's goals. His conversations with aerospace contractors around possible work for SpaceX left Musk disenchanted. It sounded like they all charged a lot of money and worked slowly. The plan to integrate components made by these types of companies gave way to the decision to make as much as practical right at SpaceX. "While drawing upon the ideas of many prior launch vehicle programs from Apollo to the X-34/Fastrac, SpaceX is privately developing the entire Falcon rocket from the ground up, including both engines, the turbo-pump, the cryogenic tank structure and the guidance system," the company announced on its website. "A ground up internal development increases difficulty and the required investment, but no other path will achieve the needed improvement in the cost of access to space."

The SpaceX executives Musk hired were an all-star crew. Mueller set to work right away building the two engines—Merlin and Kestrel, named after two types of falcons. Chris Thompson, a onetime marine who had managed the production of the Delta and Titan rockets at Boeing, joined as the vice president of operations. Tim Buzza also came from Boeing, where he'd earned a reputation as one of the world's leading rocket testers. Steve Johnson, who had worked at JPL and at two commercial space companies, was tapped as the senior mechanical engineer. The aerospace engineer Hans Koenigsmann came on to develop the avionics, guidance, and control systems. Musk also recruited Gwynne Shotwell, an aerospace veteran who started as SpaceX's first salesperson and rose in the years that followed to be president and Musk's right-hand woman.

These early days also marked the arrival of Mary Beth Brown, a now-legendary character in the lore of both SpaceX and Tesla. Brown—or MB, as everyone called her—became Musk's loyal assistant, establishing a real-life version of the relationship between *Iron Man*'s Tony Stark and Pepper Potts. If Musk worked a twenty-hour day, so too did Brown. Over the years, she brought Musk meals, set up his business appointments, arranged time with his children, picked out his clothes, dealt with press requests, and when necessary yanked Musk out of meetings to keep him on schedule. She would emerge as the only bridge between Musk and all of his interests and was an invaluable asset to the companies' employees.

Brown played a key role in developing SpaceX's early culture. She paid attention to small details like the office's red spaceship trash cans and helped balance the vibe around the office. When it came to matters related directly to Musk, Brown put on her firm countenance and no-nonsense attitude. The rest of the time she usually had a warm, broad smile and a disarming charm. "It was always, 'Oh, dear. How are you, dear?'" recalled a SpaceX technician. Brown collected the weird e-mails that arrived for Musk and sent them out as "Kook of the Week" missives to make people laugh. One of the better entries included a pencil sketch of a lunar spacecraft that had a red spot on the page. The person who sent in the letter had circled the spot on his own drawing and then written "What is that? Blood?" next to it. In other letters there were plans for a perpetual motion machine and a proposal for a giant inflatable rabbit that could be used to plug oil spills. For a short time, Brown's duties extended to managing SpaceX's books and handling the flow of business in Musk's absence. "She pretty much called the shots," the technician said. "She would say, 'This is what Elon would want.'"

Her greatest gift, though, may have been reading Musk's moods. At both SpaceX and Tesla, Brown placed her desk a few feet in front of Musk's, so that people had to pass her before having a meeting with him. If someone needed to request permission to buy a big-ticket item, they would stop for a moment in front of Brown and wait for a nod to go see Musk or the shake-off to go away because Musk was having a bad day. This system of nods and shakes became particularly important during periods of romantic strife for Musk, when his nerves were on edge more than usual.

The rank-and-file engineers at SpaceX tended to be young, male overachievers. Musk would personally reach out to the aerospace departments of top colleges and inquire about the students who had finished with the best marks on their exams. It was not unusual for him to call the students in their dorm rooms and recruit them over the phone. "I thought it was a prank call," said Michael Colonno, who heard from Musk while attending Stanford. "I did not believe for a minute that he had a rocket company." Once the students looked Musk up on the Internet, selling them on SpaceX was easy. For the first time in years if not decades, young aeronautics whizzes who pined to explore space had a really exciting company to latch on to and a path toward designing a rocket or even becoming an astronaut that did not require them to join a bureaucratic government contractor. As word of SpaceX's ambitions spread, top engineers from Boeing, Lockheed Martin, and Orbital Sciences with a high tolerance for risk fled to the upstart, too.

Throughout the first year at SpaceX, one or two new employees joined almost every week. Kevin Brogan was employee No. 23 and came from TRW, where he'd been used to various internal policies blocking him from doing work. "I called it the country club," he said. "Nobody did anything." Brogan started at SpaceX the day after his interview and was told to go hunting in the office for a computer to use. "It was go to Fry's and get whatever you need and go to Staples and get a chair," Brogan said. He immediately felt in over his head and would work for twelve hours, drive home, sleep for ten hours, and then head right back to the factory. "I was exhausted and out of shape mentally," he said. "But soon I loved it and got totally hooked."

One of the first projects SpaceX decided to tackle was the construction of a gas generator, a machine not unlike a small rocket engine that produces hot gas. Mueller, Buzza, and a couple of young engineers assembled the generator in Los Angeles and then packed it into the back of a pickup truck and drove it out to Mojave, California, to test it. A desert town about one hundred miles from Los Angeles, Mojave had become a hub for aerospace companies like Scaled Composites and XCOR. A lot of the aerospace projects were based out of the Mojave airport, where companies had their workshops and sent up all manner of cutting-edge airplanes and rockets. The SpaceX team fit right into this environment and borrowed a test stand from XCOR that was just about the perfect size to hold the gas generator. The first ignition run took place at 11 A.M. and lasted ninety seconds. The gas generator worked, but it had let out a billowing black smoke cloud that on this windless day parked right over the airport tower. The airport manager came down to the test area and lit into Mueller and Buzza. The airport official and some of the guys from XCOR who had been helping out urged the SpaceX engineers to take it easy and wait until the next day to run another test. Instead, Buzza a strong leader ready to put SpaceX's relentless ethos into play, coordinated a couple of trucks to pick up more fuel, talked the airport manager down, and got the test stand ready for another fire. In the days that followed, SpaceX's engineers perfected a routine that let them do multiple tests a day—an unheard-of practice at the airport—and had the gas generator tuned to their liking after two weeks of work.

They made a few more trips to Mojave and some other spots, including a test stand at Edwards Air Force Base and another in Mississippi. While on this countrywide rocketry tour, the SpaceX engineers came across a three-hundred-acre test site in McGregor, Texas, a small city near the center of the state. They really liked this spot, and talked Musk into buying it. The navy had tested rockets on the land years before and so too had Andrew Beal before his aerospace company collapsed. "After Beal saw it was going to cost him $300 million to develop a rocket capable of sending sizeable satellites into orbit, he called it quits, leaving behind a lot of useful infrastructure for SpaceX, including a three-story concrete tripod with legs as big around as redwood tree trunks," wrote journalist Michael Belfiore in *Rocketeers*, a book that captured the rise of a handful of private space companies.

Jeremy Hollman was one of the young engineers who soon found himself living in Texas and customizing the test site to SpaceX's needs. Hollman exemplified the kind of recruit Musk wanted: he'd earned an aerospace engineering degree from Iowa State University and a master's in astronautical engineering from the University of Southern California. He'd spent a couple of years working as a test engineer at Boeing dealing with jets, rockets, and spacecraft.*

The stint at Boeing had left Hollman unimpressed with big aerospace. His first day on the job came right as Boeing completed its merger with McDonnell Douglas. The resultant mammoth government contractor held a picnic to boost morale but ended up failing at even this simple exercise. "The head of one of the departments gave a speech about it being one company with one vision and then added that the company was very cost constrained," Hollman said. "He asked that everyone limit themselves to one piece of chicken." Things didn't improve much from

---

* Buzza knew Hollman's work at Boeing and coaxed him to SpaceX about six months after the company started.

there. Every project at Boeing felt large, cumbersome, and costly. So, when Musk came along selling radical change, Hollman bit. "I thought it was an opportunity I could not pass up," he said. At twenty-three, Hollman was young, single, and willing to give up any semblance of having a life in favor of working at SpaceX non-stop, and he became Mueller's second in command.

Mueller had developed a pair of three-dimensional computer models of the two engines he wanted to build. Merlin would be the engine for the first stage of the Falcon 1, which lifted it off the ground, and Kestrel would be the smaller engine used to power the upper, second stage of the rocket and guide it in space. Together, Hollman and Mueller figured out which parts of the engines SpaceX would build at the factory and which parts it would try to buy. For the purchased parts, Hollman had to head out to various machine shops and get quotes and delivery dates for the hardware. Quite often, the machinists told Hollman that SpaceX's timelines were nuts. Others were more accommodating and would try to bend an existing product to SpaceX's needs instead of building something from scratch. Hollman also found that creativity got him a long way. He discovered, for example, that changing the seals on some readily available car wash valves made them good enough to be used with rocket fuel.

After SpaceX completed its first engine at the factory in California, Hollman loaded it and mounds of other equipment into a U-Haul trailer. He hitched the U-Haul to the back of a white Hummer H2 and drove four thousand pounds of gear* across Interstate 10 from Los Angeles to Texas and the test site. The arrival of the engine in Texas kicked off one of the great bonding exercises in SpaceX's history. Amid rattlesnakes, fire ants, isolation, and searing heat, the group led by Buzza and Mueller began

---

* Including a 1,300-pound hunk of copper.

the process of exploring every intricacy of the engines. It was a high-pressure slog full of explosions—or what the engineers politely called "rapid unscheduled disassemblies"—that would determine whether a small band of engineers really could match the effort and skill of nation-states. The SpaceX employees christened the site in fitting fashion, downing a $1,200 bottle of Rémy Martin cognac out of paper cups and passing a sobriety test on the drive back to the company apartments in the Hummer. From that point on, the trek from California to the test site became known as the Texas Cattle Haul. The SpaceX engineers would work for ten days straight, come back to California for a weekend, and then head back. To ease the burden of travel, Musk sometimes let them use his private jet. "It carried six people," Mueller said. "Well, seven if someone sat in the toilet, which happened all the time."

While the navy and Beal had left some testing apparatus, SpaceX had to build a large amount of custom gear. One of the largest of these structures was a horizontal test stand about 30 feet long, 15 feet wide, and 15 feet tall. Then there was the complementary vertical test stand that stood two stories high. When an engine needed to be fired, it would be fastened to one of the test stands, outfitted with sensors to collect data, and monitored via several cameras. The engineers took shelter in a bunker protected on one side by a dirt embankment. If something went wrong, they would look at feeds from the webcams or slowly lift one of the bunker's hatches to listen for any clues. The locals in town rarely complained about the noise, although the animals on nearby farms seemed less impressed. "Cows have this natural defense mechanism where they gather and start running in a circle," Hollman said. "Every time we fired an engine, the cows scattered and then got in that circle with the younger ones placed in the middle. We set up a cow cam to watch them."

Both Kestrel and Merlin came with challenges, and they were treated as alternating engineering exercises. "We would run Merlin until we ran out of hardware or did something bad," Mueller said. "Then we'd run Kestrel and there was never a shortage of things to do." For months, the SpaceX engineers arrived at the site at 8 A.M. and spent twelve hours there working on the engines before retiring to the Outback Steakhouse for dinner. Mueller had a particular knack for looking over test data and spotting some place where the engine ran hot or cold or had another flaw. He would call California and prescribe hardware changes, and engineers would refashion parts and send them off to Texas. Often the workers in Texas modified parts themselves using a mill and lathe that Mueller had brought out. "Kestrel started out as a real dog, and one of my proudest moments was taking it from terrible to great performance with stuff we bought online and did in the machine shop," Mueller said. Some members of the Texas crew honed their skills to the point that they could build a test-worthy engine in three days. These same people were required to be adept at software. They'd pull an all-nighter building a turbo pump for the engine and then dig in the next night to retool a suite of applications used to control the engines. Hollman did this type of work all the time and was an all-star, but he was not alone among this group of young, nimble engineers who crossed disciplines out of necessity and the spirit of adventure. "There was an almost addictive quality to the experience," Hollman said. "You're twenty-four or twenty-five, and they're trusting you with so much. It was very empowering."

To get to space, the Merlin engine would need to burn for 180 seconds. That seemed like an eternity for the engineers at the outset of their stint in Texas, when the engine would burn for only a half second before it conked out. Sometimes Merlin vibrated too much during the tests. Sometimes it responded badly to a new material. Sometimes it cracked and needed major part upgrades, like moving from an aluminum manifold to a manifold made out of the more exotic Inconel, an alloy suited to extreme temperatures. On one occasion, a fuel valve refused to open properly and caused the whole engine to blow up. Another test gone wrong ended up with the whole test stand burning down. It usually came to Buzza and Mueller to make the unpleasant call back to Musk and recap the day's foibles. "Elon had pretty good patience," Mueller said. "I remember one time we had two test stands running and blew up two things in one day. I told Elon we could put another engine on there, but I was really, really frustrated and just tired and mad and was kinda short with Elon. I said, 'We can put another fucking thing on there, but I've blown up enough shit today.' He said, 'Okay, all right, that's fine. Just calm down. We'll do it again tomorrow.'" Coworkers in El Segundo later reported that Musk had been near tears during this call after hearing the frustration and agony in Mueller's voice.

What Musk would not tolerate were excuses or the lack of a clear plan of attack. Hollman was one of many engineers who arrived at this realization after facing one of Musk's trademark grillings. "The worst call was the first one," Hollman said. "Something had gone wrong, and Elon asked me how long it would take to be operational again, and I didn't have an immediate answer. He said, 'You need to. This is important to the company. Everything is riding on this. Why don't you have an answer?' He kept hitting me with pointed, direct questions. I thought it was more important to let him know quickly what happened, but I learned it was more important to have all the information."

From time to time, Musk participated in the testing process firsthand. One of the more memorable examples of this came as SpaceX tried to perfect a cooling chamber for its engines. The company had bought several of these chambers at $75,000 a pop and needed to put them under pressure with water to gauge their ability to handle stress. During the initial test, one of the pricey chambers cracked. Then the second one broke in the same place. Musk ordered a third test, as the engineers looked on in horror. They thought the test might be putting the chamber under undue stress and that Musk was burning through essential equipment. When the third chamber cracked, Musk flew the hardware back to California, took it to the factory floor, and, with the help of some engineers, started to fill the chambers with an epoxy to see if it would seal them. "He's not afraid to get his hands dirty," Mueller said. "He's out there with his nice Italian shoes and clothes and has epoxy all over him. They were there all night and tested it again and it broke anyway." Musk, clothes ruined, had decided the hardware was flawed, tested his hypothesis, and moved on quickly, asking the engineers to come up with a new solution.

These incidents were all part of a trying but productive process. SpaceX had developed the feeling of a small, tight-knit family up against the world. In late 2002, the company had an empty warehouse. One year later, the facility looked like a real rocket factory. Working Merlin engines were arriving back from Texas, and being fed into an assembly line where machinists could connect them to the main body, or first stage, of the rocket. More stations were set up to link the first stage with the upper stage of the rocket. Cranes were placed on the floor to handle the heavy lifting of components, and blue metal transport tracks were positioned to guide the rocket's body through the factory from station to station. SpaceX had also started to build the fairing, or case, that protects payloads atop the rocket during launch and then opens up like a clam in space to let out the cargo.

SpaceX had picked up a customer as well. According to Musk, its first rocket would launch in "early 2004" from Vandenberg Air

Force Base, carrying a satellite called TacSat-1 for the Department of Defense. With this goal looming, twelve-hour days, six days a week were considered the norm, although many people worked longer than that for extended periods of time. Respites, as far as they existed, came around 8 P.M. on some weeknights when Musk would allow everyone to use their work computers to play first-person-shooter video games like Quake III Arena and Counter-Strike against each other. At the appointed hour, the sound of guns loading would cascade throughout the office as close to twenty people armed themselves for battle. Musk—playing under the handle Random9—often won the games, talking trash and blasting away his employees without mercy. "The CEO is there shooting at us with rockets and plasma guns," said Colonno. "Worse, he's almost alarmingly good at these games and has insanely fast reactions. He knew all the tricks and how to sneak up on people."

The pending launch ignited Musk's salesman instincts. He wanted to show the public what his tireless workers had accomplished and drum up some excitement around SpaceX. Musk decided to unveil a prototype of Falcon 1 to the public in December 2003. The company would haul the seven-story-high Falcon 1 across the country on a specially built rig and leave it—and the SpaceX mobile launch system—outside of the Federal Aviation Administration's headquarters in Washington, D.C. An accompanying press conference would make it clear to Washington that a modern, smarter, cheaper rocket maker had arrived.

This marketing song and dance didn't sound sensible to SpaceX's engineers. They were working more than one hundred hours per week to make the actual rocket that SpaceX would need to be in business. Musk wanted them to do that and build a slick-looking mock-up. Engineers were called back from Texas and assigned another ulcer-inducing deadline to craft this prop. "In my mind, it was a boondoggle," Hollman said. "It wasn't advancing anything. In Elon's mind, it would get us a lot of backing from important people in the government."

While making the prototype for the event, Hollman experienced the full spectrum of highs and lows that came with working for Musk. The engineer had lost his regular glasses weeks earlier when they slipped off his face and fell down a flame duct at the Texas test site. Hollman had since made do by wearing an old pair of prescription safety glasses,* but they too were ruined when he scratched the lenses while trying to duck under an engine at the SpaceX factory. Without a spare moment to visit an optometrist, Hollman started to feel his sanity fray. The long hours, the scratch, the publicity stunt—they were all too much.

He vented about this in the factory one night, unaware that Musk stood nearby and could hear everything. Two hours later, Mary Beth Brown appeared with an appointment card to see a Lasik eye surgery specialist. When Hollman visited the doctor, he discovered that Musk had already agreed to pay for the surgery. "Elon can be very demanding, but he'll make sure the obstacles in your way are removed," Hollman said. Upon reflection, he also warmed to the long-term thinking behind Musk's Washington plan. "I think he wanted to add an element of realism to SpaceX, and if you park a rocket in someone's front yard, it's hard to deny it," Hollman said.

The event in Washington ended up being well received, and just a few weeks after it took place, SpaceX made another astonishing announcement. Despite not having even flown a rocket yet, SpaceX revealed plans for a second rocket. Along with the

---

* Before returning to El Segundo, Hollman used a drill press to remove the glasses' safety shield. "I didn't want to look like a nerd on the flight home," he said.

Falcon 1, it would build the Falcon 5. Per the name, this rocket would have five engines and could carry more weight—9,200 pounds—to low orbit around Earth. Crucially, the Falcon 5 could also theoretically reach the International Space Station for resupply missions—a capability that would open up SpaceX for some large NASA contracts. And, in a nod to Musk's obsession with safety, the rocket was said to be able to complete its missions even if three of the five engines failed, which was a level of added reliability that had not been seen in the market in decades.

The only way to keep up with all of this work was to do what SpaceX had promised from the beginning: operate in the spirit of a Silicon Valley start-up. Musk was always looking for brainy engineers who had not just done well at school but had done something exceptional with their talents. When he found someone good, Musk was relentless in courting him or her to come to SpaceX. Bryan Gardner, for example, first met Musk at a space rave in the hangars at the Mojave airport and a short while later started talking about a job. Gardner was having some of his academic work sponsored by Northrop Grumman. "Elon said, 'We'll buy them out,'" Gardner said. "So, I e-mailed him my resume at two thirty A.M., and he replied back in thirty minutes addressing everything I put in there point by point. He said, 'When you interview make sure you can talk concretely about what you do rather than use buzzwords.' It floored me that he would take the time to do this." After being hired, Gardner was tasked with improving the system for testing the valves on the Merlin engine. There were dozens of valves, and it took three to five hours to manually test each one. Six months later, Gardner had built an automated system for testing the valves in minutes. The testing machine tracked the valves individually, so that an engineer in Texas could request what the metrics had been on a specific part. "I had been handed this redheaded stepchild that no one else wanted to deal with and established my engineering credibility," Gardner said.

As the new hires arrived, SpaceX moved beyond its original building to fill up several buildings in the El Segundo complex. The engineers were running demanding software and rendering large graphics files and needed high-speed connections between all of these offices. But SpaceX had neighbors who were blocking an initiative to connect all of its buildings via fiber optic lines. Instead of taking the time to haggle with the other companies for right of way, the IT chief Branden Spikes, who had worked with Musk at Zip2 and PayPal, came up with a quicker, more devious solution. A friend of his worked for the phone company and drew a diagram that demonstrated a way to squeeze a networking cable safely between the electricity, cable, and phone wires on a telephone pole. At 2 A.M., an off-the-books crew showed up with a cherry picker and ran fiber to the telephone poles and then ran cables straight to the SpaceX buildings. "We did that over a weekend instead of taking months to get permits," Spikes said. "There was always this feeling that we were facing a sort of insurmountable challenge and that we had to band together to fight the good fight." SpaceX's landlord, Alex Lidow, chuckled when thinking back to all of the antics of Musk's team. "I know they did a lot of hanky stuff at night," he said. "They were smart, needed to get things done, and didn't always have time to wait for things like city permits."

Musk never relented in asking his employees to do more and be better, whether it was at the office or during extracurricular activities. Part of Spikes's duties included building custom gaming PCs for Musk's home that pushed their computational power to the limits and needed to be cooled with water running through a series of tubes inside the machines. When one of these gaming rigs kept breaking, Spikes figured out that Musk's mansion had dirty power lines and had a second, dedicated power circuit built for the gaming room to correct the problem. Doing this favor bought Spikes no special treatment. "SpaceX's mail server crashed one time, and Elon word for word said, 'Don't ever fucking let that happen again,'" Spikes said. "He had a way of looking at you—a glare—and would keep looking at you until you understood him."

Musk had tried to find contractors that could keep up with SpaceX's creativity and pace. Instead of always hitting up aerospace guys, for example, he located suppliers with similar experience from different fields. Early on, SpaceX needed someone to build the fuel tanks, essentially the main body of the rocket, and Musk ended up in the Midwest talking to companies that had made large, metal agricultural tanks used in the dairy and food processing businesses. These suppliers also struggled to keep up with SpaceX's schedule, and Musk found himself flying across the country to pay visits—sometimes surprise ones—on the contractors to check on their progress. One such inspection took place at a company in Wisconsin called Spincraft. Musk and a couple of SpaceX employees flew his jet across the country and arrived late at night expecting to see a shift of workers doing extra duty to get the fuel tanks completed. When Musk discovered that Spincraft was well behind schedule, he turned to a Spincraft employee and informed him, "You're fucking us up the ass, and it doesn't feel good." David Schmitz was a general manager at Spincraft and said Musk earned a reputation as a fearsome negotiator who did indeed follow up on things personally. "If Elon was not happy, you knew it," Schmitz said. "Things could get nasty." In the months that followed that meeting, SpaceX increased its internal welding capabilities so that it could make the fuel tanks in El Segundo and ditch Spincraft.

Another salesman flew down to SpaceX to sell the company on some technology infrastructure equipment. He was doing the standard relationship-building exercise practiced by salespeople for centuries. Show up. Speak for a while. Feel each other out. Then, start doing business down the road. Musk was having none of it. "The guy comes in, and Elon asks him why they're meeting," Spikes said. "He said, 'To develop a relationship.' Elon replied, 'Okay. Nice to meet you,' which basically meant, 'Get the fuck out of my office.' This guy had spent four hours traveling for what ended up as a two-minute meeting. Elon just has no tolerance for that kind of stuff." Musk could be equally brisk with employees who were not hitting his standards. "He would often say, 'The longer you wait to fire someone the longer it has been since you should have fired them,'" Spikes said.

Most of the SpaceX employees were thrilled to be part of the company's adventure and tried not to let Musk's grueling demands and harsh behavior get to them. But there were some moments where Musk went too far. The engineering corps flew into a collective rage every time they caught Musk in the press claiming to have designed the Falcon rocket more or less by himself. Musk also hired a documentary crew to follow him around for a while. This audacious gesture really grated on the people toiling away in the SpaceX factory. They felt like Musk's ego had gotten the best of him and that he was presenting SpaceX as the conqueror of the aerospace industry when the company had yet to launch successfully. Employees who made detailed cases around what they saw as flaws in the Falcon 5 design or presented practical suggestions to get the Falcon 1 out the door more quickly were often ignored or worse. "The treatment of staff was not good for long stretches of this era," said one engineer. "Many good engineers, who everyone beside 'management' felt were assets to the company, were forced out or simply fired outright after being blamed for things they hadn't done. The kiss of death was proving Elon wrong about something."

Early 2004, when SpaceX had hoped to launch its rocket, came and went. The Merlin engine that Mueller and his team had built appeared to be among the most efficient rocket engines ever made. It was just taking longer than Musk had expected to pass tests needed to clear the engine for a launch. Finally, in the fall of 2004, the engines were burning consistently and meeting all their requirements. This meant that Mueller and his team could breathe easy and that everyone else at SpaceX should prepare to suffer. Mueller had spent SpaceX's entire existence as the "critical path"—the person holding up the company from achieving its next steps—working under Musk's scrutiny. "With the engine ready, it was time for mass panic," Mueller said. "No one else knew what it was like to be on critical path."

Lots of people soon found out, as major problems abounded. The avionics, which included the electronics for the navigation, communication, and overall management of the rocket, turned into a nightmare. Seemingly trivial things like getting a flash storage drive to talk to the rocket's main computer failed for undetectable reasons. The software needed to manage the rocket also became a major burden. "It's like anything else where you find out that the last ten percent is where all the integration happens and things don't play together," Mueller said. "This process went on for six months." Finally, in May 2005, SpaceX transported the rocket 180 miles north to Vandenberg Air Force Base for a test fire and completed a five-second burn on the launchpad.

Launching from Vandenberg would have been very convenient for SpaceX. The site is close to Los Angeles and has several launchpads to pick from. SpaceX, though, became an unwelcome guest. The air force gave the newcomer a cool welcome, and the people assigned to manage the launch sites did not go

out of their way help SpaceX. Lockheed and Boeing, which fly
$1 billion spy satellites for the military from Vandenberg, didn't
care for SpaceX's presence, either—in part because SpaceX represented a threat to their business and in part because this start-
up was mucking around near their precious cargo. As SpaceX
started to move from the testing phase to the launch, it was told
to get in line. They would have to wait months to launch. "Even
though they said we could fly, it was clear that we would not," said
Gwynne Shotwell.

Searching for a new site, Shotwell and Hans Koenigsmann
put a Mercator projection of the world up on the wall and looked
for a name they recognized along the equator, where the planet
spins faster and gives rockets an added boost. The first name that
jumped out was Kwajalein Island—or Kwaj—the largest island in
an atoll between Guam and Hawaii in the Pacific Ocean and part
of the Republic of the Marshall Islands. This spot registered with
Shotwell because the U.S. Army had used it for decades as a missile test site. Shotwell looked up the name of a colonel at the test
site and sent him an e-mail, and three weeks later got a call back
with the army saying they would love to have SpaceX fly from the
islands. In June 2005, SpaceX's engineers began to fill containers
with their equipment to ship them to Kwaj.

About one hundred islands make up the Kwajalein Atoll.
Many of them stretch for just a few hundred yards and are much
longer than they are wide. "From the air, the place looks like
these beautiful beads on a string," said Pete Worden, who visited
the site in his capacity as a Defense Department consultant. Most
of the people in the area live on an island called Ebeye, while the
U.S. military has taken over Kwajalein, the southernmost island,
and turned it into part tropical paradise and part Dr. Evil's secret
lair. The United States spent years lobbing its ICBMs from California at Kwaj and used the island to run experiments on its space

weapons during the "Star Wars" period. Laser beams would be aimed at Kwaj from space in a bid to see if they were accurate and responsive enough to take out an ICBM hurtling toward the islands. The military presence resulted in a weird array of buildings including hulking, windowless trapezoidal concrete structures clearly conceived by someone who deals with death for a living.

To get to Kwaj, the SpaceX employees either flew on Musk's jet or took commercial flights through Hawaii. The main accommodations were two-bedroom affairs on Kwajalein Island that looked more like dormitories than hotel rooms, with their military-issued dressers and desks. Any materials that the engineers needed had to be flown in on Musk's plane or were more often brought by boat from Hawaii or the mainland United States. Each day, the SpaceX crew gathered their gear and took a forty-five-minute boat ride to Omelek, a seven-acre, palm-tree-and-vegetation-covered island that would be transformed into their launchpad. Over the course of several months, a small team of people cleared the brush, poured concrete to support the launchpad, and converted a double-wide trailer into offices. The work was grueling and took place in soul-sapping humidity under a sun powerful enough to burn the skin through a T-shirt. Eventually, some of the workers preferred to spend the night on Omelek rather than make the journey through rough waters back to the main island. "Some of the offices were turned into bedrooms with mattresses and cots," Hollman said. "Then we shipped over a very nice refrigerator and a good grill and plumbed in a shower. We tried to make it less like camping and more like living."

The sun rose at 7 A.M. each day, and that's when the SpaceX team got to work. A series of meetings would take place with people listing what needed to get done, and debating solutions to lingering problems. As the large structures arrived, the workers placed the body of the rocket horizontally in a makeshift hangar and spent hours melding together all of its parts. "There was always something to do," Hollman said. "If the engine wasn't a problem, then there was an avionics problem or a software problem." By 7 P.M., the engineers wound down their work. "One or two people would decide it was their night to cook, and they would make steak and potatoes and pasta," Hollman said. "We had a bunch of movies and a DVD player, and some of us did a lot of fishing off the docks." For many of the engineers, this was both a torturous and magical experience. "At Boeing you could be comfortable, but that wasn't going to happen at SpaceX," said Walter Sims, a SpaceX tech expert who found time to get certified to dive while on Kwaj. "Every person on that island was a fucking star, and they were always holding seminars on radios or the engine. It was such an invigorating place."

The engineers were constantly baffled by what Musk would fund and what he wouldn't. Back at headquarters, someone would ask to buy a $200,000 machine or a pricey part that they deemed essential to Falcon 1's success, and Musk would deny the request. And yet he was totally comfortable paying a similar amount to put a shiny surface on the factory floor to make it look nice. On Omelek, the workers wanted to pave a two-hundred-yard pathway between the hangar and the launchpad to make it easier to transport the rocket. Musk refused. This left the engineers moving the rocket and its wheeled support structure in the fashion of the ancient Egyptians. They laid down a series of wooden planks and rolled the rocket across them, grabbing the last piece of wood from the back and running it forward in a continuous cycle.

The whole situation was ludicrous. A start-up rocket company had ended up in the middle of nowhere trying to pull off one of the most difficult feats known to man, and, truth be told, only a handful of the SpaceX team had any idea how to make a launch happen. Time and again, the rocket would get marched out to the launchpad and hoisted vertical for a couple of days, while technical and safety checks would reveal a litany of new problems. The engineers worked on the rocket for as long as they could before laying it horizontal and marching it back to the hangar to avoid damage from the salty air. Teams that had worked separately for months back at the SpaceX factory—propulsion, avionics, software—were thrust together on the island and forced to become an interdisciplinary whole. The sum total was an extreme learning and bonding exercise that played like a comedy of errors. "It was like *Gilligan's Island* except with rockets," Hollman said.

In November 2005, about six months after they had first gotten to the island, the SpaceX team felt ready to give launching a shot. Musk flew in with his brother, Kimbal, and joined the majority of the SpaceX team in the barracks on Kwaj. On November 26, a handful of people woke up at 3 A.M. and filled the rocket with liquid oxygen. They then scampered off to an island about three miles away for protection, while the rest of the SpaceX team monitored the launch systems from a control room twenty-six miles away on Kwaj. The military gave SpaceX a six-hour launch window. Everyone was hoping to see the first stage take off and reach about 6,850 miles per hour before giving way to the second stage, which would ignite up in the air and reach 17,000 miles per hour. But, while going through the pre-launch checks, the engineers detected a major problem: a valve on a liquid oxygen tank would not close, and the LOX was boiling off into the air at 500 gallons per hour. SpaceX scrambled to fix the issue but lost too much of its fuel to launch before the window closed.

With that mission aborted, SpaceX ordered major LOX reinforcements from Hawaii and prepared for another attempt in mid-December. High winds, faulty valves, and other errors thwarted that launch attempt. Before another attempt could be made, SpaceX discovered on a Saturday night that the rocket's power distribution systems had started malfunctioning and would need new capacitors. On Sunday morning, the rocket was lowered and split into its two stages so that a technician could slide in and remove the electrical boards. Someone found an electronics supplier that was open on Sunday in Minnesota, and off a SpaceX employee flew to get some fresh capacitors. By Monday he was in California and testing the parts at SpaceX's headquarters to make sure they passed various heat and vibration checks, then on a plane again back to the islands. In under eighty hours, the electronics had been returned in working order and installed in the rocket. The dash to the United States and back showed that SpaceX's thirty-person team had real pluck in the face of adversity and inspired everyone on the island. A traditional three-hundred-person-strong aerospace launch crew would never have tried to fix a rocket like that on the fly. But the energy, smarts, and resourcefulness of the SpaceX team still could not overcome their inexperience or the difficult conditions. More problems arose and blocked any thoughts of a launch.

Finally, on March 24, 2006, it was all systems go. The Falcon 1 stood on its square launchpad and ignited. It soared into the sky, turning the island below it into a green spec amid a vast, blue expanse. In the control room, Musk paced as he watched the action, wearing shorts, flip-flops, and a T-shirt. Then, about twenty-five seconds in, it became clear that all was not well. A fire broke out above the Merlin engine and suddenly this machine that had been flying straight and true started to spin and then tumble uncontrollably back to Earth. The Falcon 1 ended up falling directly down onto the launch site. Most of the debris went into a reef 250 feet from the launchpad, and the satellite cargo smashed through SpaceX's machine shop roof and landed more or less intact on the floor. Some of the engineers put on their snorkeling and scuba gear and recovered the pieces, fitting all of the rocket's remnants into two refrigerator-sized crates. "It is perhaps worth noting that those launch companies that succeeded also took their lumps along the way," Musk wrote in a postmortem. "A friend of mine wrote to remind me that only 5 of the first 9 Pegasus launches succeeded; 3 of 5 for Ariane; 9 of 20 for Atlas; 9 of 21 for Soyuz; and 9 of 18 for Proton. Having experienced firsthand how hard it is to reach orbit, I have a lot of respect for those that persevered to produce the vehicles that are mainstays of space launch today." Musk closed the letter writing, "SpaceX is in this for the long haul and, come hell or high water, we are going to make this work."

Musk and other SpaceX executives blamed the crash on an unnamed technician. They said this technician had done some work on the rocket one day before the launch and failed to properly tighten a fitting on a fuel pipe, which caused the fitting to crack. The fitting in question was something basic—an aluminum b-nut that's often used to connect a pair of tubes. The technician was Hollman. In the aftermath of the rocket crash, Hollman flew to Los Angeles to confront Musk directly. He'd spent years working day and night on the Falcon 1 and felt enraged that Musk had called out him and his team in public. Hollman knew that he'd fastened the b-nut correctly and that observers from NASA had been looking over his shoulder to check the work. When Hollman charged into SpaceX's headquarters with a head full of fury, Mary Beth Brown tried to calm him and stop him from seeing Musk. Hollman kept going anyway, and the two of them proceeded to have a shouting match at Musk's cubicle.

After all the debris was analyzed, it turned out that the b-nut had almost certainly cracked due to corrosion from the months in Kwaj's salty atmosphere. "The rocket was literally crusted with salt on one side, and you had to scrape it off," Mueller said. "But we had done a static fire three days earlier, and everything was fine." SpaceX had tried to save about fifty pounds of weight by using aluminum components instead of stainless steel. Thompson, the former marine, had seen the aluminum parts work just fine in helicopters that sat on aircraft carriers, and Mueller had seen aircraft resting outside of Cape Canaveral for forty years with aluminum b-nuts in fine condition. Years later, a number of SpaceX's executives still agonize over the way Hollman and his team were treated. "They were our best guys, and they kind of got blamed to get an answer out to the world," Mueller said. "That was really bad. We found out later that it was dumb luck."*

After the crash, there was a lot of drinking at a bar on the main island. Musk wanted to launch again within six months, but putting together a new machine would again require an immense amount of work. SpaceX had some pieces for the vehicle ready in El Segundo but certainly not a ready-to-fire rocket. As they downed drinks, the engineers vowed to take a more disciplined approach with their next craft and to work better as a collective. Worden hoped the SpaceX engineers would raise their game as well. He'd been observing them for the Defense Department and loved the energy of the young engineers but not their methodology. "It was being done like a bunch of kids in Silicon Valley would do software," Worden said. "They would stay up all night and try this and try that. I'd seen hundreds of these types of operations, and it struck me that it wouldn't work." Leading up to the

---

*   Hollman left the company after this incident in November 2007 and then returned for a spell to train new personnel. A number of people I interviewed for the book said that Hollman was so key to SpaceX's early days that they feared the company might flame out without him.

first launch, Worden tried to caution Musk, sending a letter to him and the director of DARPA, the research arm of the Defense Department, that made his views clear. "Elon didn't react well. He said, 'What do you know? You're just an astronomer,'" Worden said. But, after the rocket blew up, Musk recommended that Worden perform an investigation for the government. "I give Elon huge credit for that," Worden said.

Almost exactly a year later, SpaceX was ready to try another launch. On March 15, 2007, a successful test fire took place. Then, on March 21, the Falcon 1 finally behaved. From its launchpad surrounded by palm trees, the Falcon 1 surged up and toward space. It flew for a couple of minutes with engineers now and again reporting that the systems were "nominal," or in good shape. At three minutes into the flight, the first stage of the rocket separated and fell back to Earth, and the Kestrel engine kicked in as planned to carry the second stage into orbit. Ecstatic cheers went out in the control room. Next, at the four-minute mark, the fairing atop the rocket separated as planned. "It was doing exactly what it was supposed to do," said Mueller. "I was sitting next to Elon and looked at him and said, 'We've made it.' We're hugging and believe it's going to make it to orbit. Then, it starts to wiggle." For more than five glorious minutes, the SpaceX engineers got to feel like they had done everything right. A camera on board the Falcon 1 pointed down and showed Earth getting smaller and smaller as the rocket made its way methodically into space. But then that wiggle that Mueller noticed turned into flailing, and the machine swooned, started to break apart, and then blew up. This time the SpaceX engineers were quick to figure out what went wrong. As the propellant was consumed, what was left started to move around the tank and slosh against the sides, much like wine spinning around a glass. The sloshing propellant triggered the wobbling, and at one point it sloshed enough to leave an opening to the engine exposed. When the engine sucked in a big breath of air, it flamed out.

The failure was another crushing blow to SpaceX's engineers. Some of them had spent close to two years shuffling back and forth between California, Hawaii, and Kwaj. By the time SpaceX could attempt another launch, it would be about four years after Musk's original target, and the company had been chewing through his Internet fortune at a worrying rate. Musk had vowed publicly that he would see this thing through to the end, but people inside and outside the company were doing back-of-the-envelope math and could tell that SpaceX likely could only afford one more attempt—maybe two. To the extent that the financial situation unnerved Musk, he rarely if ever let it show to employees. "Elon did a great job of not burdening people with those worries," said Spikes. "He always communicated the importance of being lean and of success, but it was never 'if we fail, we're done for.' He was very optimistic."

The failures seemed to do little to curtail Musk's vision for the future or raise doubts about his capabilities. In the midst of the chaos, he took a tour of the islands with Worden. Musk began thinking aloud about how the islands could be unified into one landmass. He suggested that walls could be built through the small channels between the islands, and the water could be pumped out in the spirit of the manmade systems in the Netherlands. Worden, also known for his out-there ideas, was attracted to Musk's bravado. "That he is thinking of this stuff is kind of cool," Worden said. "From that point on, he and I discussed settling Mars. It really impressed me that this is a guy that thinks big."

7

ALL ELECTRIC

J. B. STRAUBEL HAS A TWO-INCH-LONG SCAR THAT CUTS ACROSS THE middle of his left cheek. He earned it in high school, during a chemistry class experiment. Straubel whipped up the wrong concoction of chemicals, and the beaker he was holding exploded, throwing off shards of glass, one of which sliced through his face.

The wound lingers as a tinkerer's badge of honor. It arrived near the end of a childhood full of experimentation with chemicals and machines. Born in Wisconsin, Straubel constructed a large chemistry lab in the basement of his family's home that included fume hoods and chemicals ordered, borrowed, or pilfered. At thirteen, Straubel found an old golf cart at the dump. He brought it back home and restored it to working condition, which required him to rebuild the electric motor. It seemed that Straubel was always taking something apart, sprucing it up, and putting it back together. All of this fit into the Straubel family's do-it-yourself traditions. In the late 1890s Straubel's great-grandfather started the Straubel Machine Company, which built one of the first internal combustion engines in the United States and used it to power boats.

Straubel's inquisitive spirit carried him west to Stanford University, where he enrolled in 1994 intending to become a physicist. After flying through the hardest courses he could take, Straubel concluded that majoring in physics would not be for him. The advanced courses were too theoretical, and Straubel liked to get his hands dirty. He developed his own major called energy systems and engineering. "I wanted to take software and electricity and use it to control energy," Straubel said. "It was computing combined with power electronics. I collected all the things I love doing in one place."

There was no clean-technology movement at this time, but there were companies dabbling with new uses for solar power and electric vehicles. Straubel ended up hunting down these start-ups, hanging out in their garages and pestering the engineers. He began tinkering once again on his own as well in the garage of a house he shared with a half dozen friends. Straubel bought a "piece of shit Porsche" for $1,600 and turned it into an electric car. This meant that Straubel had to create a controller to manage the electric motor, build a charger from scratch, and write the software that made the entire machine work. The car set the world record for electric vehicle (EV) acceleration, traveling a quarter mile in 17.28 seconds. "The thing I took away was that the electronics were great, and you could get acceleration on a shoestring budget, but the batteries sucked," Straubel said. "It had a thirty-mile range, so I learned firsthand about some of the limitations of electric vehicles." Straubel gave his car a hybrid boost, building a gasoline-powered contraption that could be towed behind the Porsche and used to recharge the batteries. It was good enough for Straubel to drive the four hundred miles down to Los Angeles and back.

By 2002, Straubel was living in Los Angeles. He'd gotten a master's degree from Stanford and bounced around a couple of companies looking for something that called out to him. He decided on Rosen Motors, which had built one of the world's first hybrid vehicles—a car that ran off a flywheel and a gas turbine and had electric motors to drive the wheel. After it folded, Straubel followed Harold Rosen, an engineer famed for inventing the geostationary satellite, to create an electric plane. "I'm a pilot and love to fly, so this was perfect for me," Straubel said. "The idea was that it would stay aloft for two weeks at a time and hover over a specific spot. This was way before drones and all that." To help make ends meet, Straubel also worked nights and on the weekend doing electronics consulting for a start-up.

It was in the midst of toiling away on all these projects that Straubel's old buddies from the Stanford solar car team came to pay him a visit. A group of rogue engineers at Stanford had been working on solar cars for years, building them in a World War II–era Quonset hut full of toxic chemicals and black widows. Unlike today, when the university would jump at the chance to support such a project, Stanford tried to shut down this group of fringe freaks and geeks. The students proved very capable of doing the work on their own and competed in cross-country solar-powered car races. Straubel helped build the vehicles during his time at university and even after, forming relationships with the incoming crop of engineers. The team had just raced 2,300 miles from Chicago to Los Angeles, and Straubel offered the strapped, exhausted kids a place to stay. About a half dozen students showed up at Straubel's place, took their first showers in many days, and then spread across his floor. As they chatted late into the night, Straubel and the solar team kept fixating on one topic. They realized that lithium ion batteries—such as the ones in their car being fed by the sun—had gotten much better than most people realized. Many consumer electronics devices like laptops were running on so-called 18650 lithium ion batteries, which looked a lot like AA batteries and could be strung together. "We wondered what would happen if you put ten thousand of the battery cells together," Straubel said. "We did the math and figured you could go almost one thousand miles. It was totally nerdy shit, and eventually everyone fell asleep, but the idea really stuck with me."

Soon enough, Straubel was stalking the solar car crew, trying to talk them into building an electric car based on the lithium ion batteries. He would fly up to Palo Alto, spend the night sleeping in his plane, and then ride a bicycle to the Stanford campus to make his sales pitch while helping with their current projects. The design Straubel had come up with was a super-aerodynamic vehicle with 80 percent of its mass made up of the batteries. It looked quite a bit like a torpedo on wheels. No one knew the exact details of Straubel's long-term vision for this thing, including Straubel. The plan seemed to be less about forming a car company than about building a proof-of-concept vehicle just to get people thinking about the power of the lithium ion batteries. With any luck, they would find a race to compete in.

The Stanford students agreed to join Straubel, if he could raise some money. He began going to trade shows handing out brochures about his idea and e-mailing just about anyone he could think of. "I was shameless," he said. The only problem was that no one had any interest in what Straubel was selling. Investors dealt him one rejection after another for months on end. Then, in the fall of 2003, Straubel met Elon Musk.

Harold Rosen had set up a lunch with Musk at a seafood restaurant near the SpaceX headquarters in Los Angeles and brought Straubel along to help talk up the electric plane idea. When Musk didn't bite on that, Straubel announced his electric car side project. The crazy idea struck an immediate chord with Musk, who had been thinking about electric vehicles for years. While Musk had mostly focused on using ultracapacitors for the vehicles, he was thrilled and surprised to hear how far the lithium ion battery technology had progressed. "Everyone else had told me I was nuts, but Elon loved the idea," Straubel said. "He said, 'Sure, I will give you some money.'" Musk promised Straubel $10,000 of the $100,000 he was seeking. On the spot, Musk and Straubel formed a kinship that would survive more than a decade of extreme highs and lows as they set out to do nothing less than change the world.

After the meeting with Musk, Straubel reached out to his friends at AC Propulsion. The Los Angeles–based company started in 1992 and was the bleeding edge of electric vehicles, building everything from zippy midsize passenger jobs right on up to sports cars. Straubel really wanted to show Musk the tzero (from "t-zero")—the highest-end vehicle in AC Propulsion's stable. It was a type of kit car that had a fiberglass body sitting on top of a steel frame and went from zero to 60 miles per hour in 4.9 seconds when first unveiled in 1997. Straubel had spent years hanging out with the AC Propulsion crew and asked Tom Gage, the company's president, to bring a tzero over for Musk to drive. Musk fell for the car. He saw its potential as a screaming-fast machine that could shift the perception of electric cars from boring and plodding to something aspirational. For months Musk offered to fund an effort to transform the kit car into a commercial vehicle but got rebuffed time and again. "It was a proof of concept and needed to be made real," Straubel said. "I love the hell out of the AC Propulsion guys, but they were sort of hopeless at business and refused to do it. They kept trying to sell Elon on this car called the eBox that looked like shit, didn't have good performance, and was just uninspiring." While the meetings with AC Propulsion didn't result in a deal, they had solidified Musk's interest in backing something well beyond Straubel's science project. In a late February 2004 e-mail to Gage, Musk wrote, "What I'm going to do is figure out the best choice of a high performance base car and electric powertrain and go in that direction."

Unbeknownst to Straubel, at about the same time, a couple of business partners in Northern California had also fallen in love with the idea of making a lithium ion battery powered car. Martin Eberhard and Marc Tarpenning had founded NuvoMedia in 1997 to create one of the earliest electronic book readers, called the Rocket eBook. The work at NuvoMedia had given the men insight into cutting-edge consumer electronics and the hugely improved lithium ion batteries used to power laptops and other portable devices. While the Rocket eBook was too far ahead of its time and not a major commercial success, it was innovative enough to attract the attention of Gemstar International Group, which owned *TV Guide* and some electronic programming guide technology. Gemstar paid $187 million to acquire NuvoMedia in March 2000. Spoils in hand, the cofounders stayed in touch after the deal. They both lived in Woodside, one of the wealthiest towns in Silicon Valley, and chatted from time to time about what they should tackle next. "We thought up some goofball things," said Tarpenning. "There was one plan for these fancy irrigation systems for farms and the home based on smart water-sensing networks. But nothing really resonated, and we wanted something more important."

Eberhard was a supremely talented engineer with a do-gooder's social conscience. The United States' repeated conflicts in the Middle East bothered him, and like many other science-minded folks around 2000 he had started to accept global warming as a reality. Eberhard began looking for alternatives to gas-guzzling cars. He investigated the potential of hydrogen fuel cells but found them lacking. He also didn't see much point in leasing something like the EV1 electric car from General Motors. What did catch Eberhard's interest, however, were the all-electric cars from AC Propulsion that he spied on the Internet. Eberhard went down to Los Angeles around 2001 to visit the AC Propulsion shop. "The place looked like a ghost town and like they were going out of business," Eberhard said. "I bailed them out with five hundred thousand dollars so that they could build one of their cars for me with lithium ion instead of lead acid batteries." Eberhard too tried to goad AC Propulsion into being a commercial enterprise rather than what was then more of a hobby shop. When they rejected his overtures, Eberhard decided to form his own company and see what the lithium ion batteries could really do.

Eberhard's journey began with him building a technical model of the electric car on a spreadsheet. This let him tweak various components and see how they might affect the vehicle's shape and performance. He could adjust the weight, number of batteries, resistance of the tires and body, and then get back answers on how many batteries it would take to power the various designs. The models made it clear that SUVs, which were very popular at the time, and things like delivery trucks were unlikely candidates. The technology seemed instead to favor a lighter-weight, high-end sports car, which would be fast, fun to drive, and have far better range than most people would expect. These technical specifications complemented the findings of Tarpenning, who had been doing research into a financial model for the car. The Toyota Prius had started to take off in California, and it was being purchased by wealthy eco-crusaders. "We also learned that the average income for EV1 owners was around two hundred thousand dollars per year," Tarpenning said. People who used to go after the Lexus, BMW, and Cadillac brands saw electric and hybrid cars as a different kind of status symbol. The men figured they could build something for the $3 billion per year luxury auto market in the United States that would let rich people have fun and feel good about themselves too. "People pay for cool and sexy and an amazing zero-to-sixty time," Tarpenning said.

On July 1, 2003, Eberhard and Tarpenning incorporated their new company. While at Disneyland a few months earlier on a date with his wife, Eberhard had come up with the name Tesla Motors, both to pay homage to the inventor and electric motor pioneer Nikola Tesla and because it sounded cool. The cofounders rented an office that had three desks and two small rooms in a decrepit 1960s building located at 845 Oak Grove Avenue in Menlo Park. The third desk was occupied a few months later by Ian Wright, an engineer who grew up on a farm in New Zealand. He was a neighbor of the Tesla cofounders in Woodside, and had been working with them to hone his pitch for a networking start-up. When the start-up failed to raise any money from venture capitalists, Wright joined Tesla. As the three men began to tell some of their confidants of their plans, they were confronted with universal derision. "We met a friend at this Woodside pub to tell her what we had finally decided to do and that it was going to be an electric car," Tarpenning said. "She said, 'You have to be kidding me.'"

Anyone who tries to build a car company in the United States is quickly reminded that the last successful start-up in the industry was Chrysler, founded in 1925. Designing and building a car from the ground up comes with plenty of challenges, but it's really getting the money and know-how to build lots of cars that has thwarted past efforts to get a new company going. The Tesla founders were aware of these realities. They figured that Nikola Tesla had built an electric motor a century earlier and that creating a drivetrain to take the power from the motor and send it to the wheels was doable. The really frightening part of their enter-

prise would be building the factory to make the car and its associated parts. But the more the Tesla guys researched the industry,
the more they realized that the big automakers don't even really
build their cars anymore. The days of Henry Ford having raw
materials delivered to one end of his Michigan factory and then
sending cars out the other end had long passed. "BMW didn't
make its windshields or upholstery or rearview mirrors," Tarpenning said. "The only thing the big car companies had kept was
internal combustion research, sales and marketing, and the final
assembly. We thought naïvely that we could access all the same
suppliers for our parts."

The plan the Tesla cofounders came up with was to license
some technology from AC Propulsion around the tzero vehicle
and to use the Lotus Elise chassis for the body of their car. Lotus,
the English carmaker, had released the two-door Elise in 1996,
and it certainly had the sleek, ground-hugging appeal to make a
statement to high-end car buyers. After talking to a number of
people in the car dealership business, the Tesla team decided to
avoid selling their cars through partners and sell direct. With
these basics of a plan in place, the three men went hunting for
some venture capital funding in January 2004.

To make things feel more real for the investors, the Tesla
founders borrowed a tzero from AC Propulsion and drove it to
the venture capital corridor of Sand Hill Road. The car acceler-
ated faster than a Ferrari, and this translated into visceral excitement for the investors. The downside, though, was that venture
capitalists are not a terribly imaginative bunch, and they struggled to see past the crappy plastic finish of this glorified kit car.
The only venture capitalists that bit were Compass Technology
Partners and SDL Ventures, and they didn't sound altogether
thrilled. The lead partner at Compass had made out well on
NuvoMedia and felt some loyalty to Eberhard and Tarpenning.

"He said, 'This is stupid, but I have invested in every automotive start-up for the last forty years, so why not,'" Tarpenning recalled. Tesla still needed a lead investor who would pony up the bulk of the $7 million needed to make what's known as a mule or a prototype vehicle. That would be their first milestone and give them something physical to show off, which could aid a second round of funding.

Eberhard and Tarpenning had Elon Musk's name in the back of their heads as a possible lead investor from the outset. They had both seen him speak a couple of years earlier at a Mars Society conference held at Stanford where Musk had laid out his vision of sending mice into space, and they got the impression that he thought a bit differently and would be open to the idea of an electric car. The idea to pitch Musk on Tesla Motors solidified when Tom Gage from AC Propulsion called Eberhard and told him that Musk was looking to fund something in the electric car arena. Eberhard and Wright flew down to Los Angeles and met with Musk on a Friday. That weekend, Musk peppered Tarpenning, who had been away on a trip, with questions about the financial model. "I just remember responding, responding, and responding," Tarpenning said. "The following Monday, Martin and I flew down to meet him again, and he said, 'Okay, I'm in.'"

The Tesla founders felt like they had lucked into the perfect investor. Musk had the engineering smarts to know what they were building. He also shared their larger goal of trying to end the United States' addiction to oil. "You need angel investors to have some belief, and it wasn't a purely financial transaction for him," Tarpenning said. "He wanted to change the energy equation of the country." With an investment of $6.5 million, Musk had become the largest shareholder of Tesla and the chairman of the company. Musk would later wield his position of strength well while battling Eberhard for control of Tesla. "It was a mistake,"

Eberhard said. "I wanted more investors. But, if I had to do it again, I would take his money. A bird in the hand, you know. We needed it."

Not long after this meeting took place, Musk called Straubel and urged him to meet with the Tesla team. Straubel heard that their offices in Menlo Park were about a half a mile from his house, and he was intrigued but very skeptical of their story. No one on the planet was more dialed into the electric vehicle scene than Straubel, and he found it hard to believe that a couple of guys had gotten this far along without word of their project reaching him. Nonetheless, Straubel stopped by the office for a meeting, and was hired right away in May 2004 at a salary of $95,000 per year. "I told them that I had been building the battery pack they need down the street with funding from Elon," Straubel said. "We agreed to join forces and formed this ragtag group."

Had anyone from Detroit stopped by Tesla Motors at this point, they would have ended up in hysterics. The sum total of the company's automotive expertise was that a couple of the guys at Tesla really liked cars and another one had created a series of science fair projects based on technology that the automotive industry considered ridiculous. What's more, the founding team had no intention of turning to Detroit for advice on how to build a car company. No, Tesla would do what every other Silicon Valley start-up had done before it, which was hire a bunch of young, hungry engineers and figure things out as they went along. Never mind that the Bay Area had no real history of this model ever having worked for something like a car and that building a complex, physical object had little in common with writing a software application. What Tesla did have, ahead of anyone else, was the realization that 18650 lithium ion batteries had gotten really good and were going to keep getting better. Hopefully that coupled with some effort and smarts would be enough.

Straubel had a direct pipeline into the smart, energetic engineers at Stanford and told them about Tesla. Gene Berdichevsky, one of the members of the solar-powered-car team, lit up the second he heard from Straubel. An undergraduate, Berdichevsky volunteered to quit school, work for free, and sweep the floors at Tesla if that's what it took to get a job. The founders were impressed with his spirit and hired Berdichevsky after one meeting. This left Berdichevsky in the uncomfortable position of calling his Russian immigrant parents, a pair of nuclear submarine engineers, to tell them that he was giving up on Stanford to join an electric car start-up. As employee No. 7, he spent part of the workday in the Menlo Park office and the rest in Straubel's living room designing three-dimensional models of the car's powertrain on a computer and building battery pack prototypes in the garage. "Only now do I realize how insane it was," Berdichevsky said.

Tesla soon needed to expand to accommodate its budding engineer army and to create a workshop that would help bring the Roadster, as they were now calling the car, to life. They found a two-story industrial building in San Carlos at 1050 Commercial Street. The 10,000-square-foot facility wasn't much, but it had room to build a research and development shop capable of knocking out some prototype cars. There were a couple of large assembly bays on the ride side of the building and two large rollup doors big enough for cars to drive in and out. Wright divided the open floor space into segments—motors, batteries, power electronics, and final assembly. The left half of the building was an office space that had been modified in weird ways by the previous tenant, a plumbing supply company. The main conference room had a wet bar and a sink where the faucet was a swan's mouth, and the hot and cold knobs were wings. Berdichevsky painted the office white on a Sunday night, and the next week the employees made a field trip to IKEA to buy desks and hopped online to order their computers from Dell. As for tools, Tesla had a single Craftsman toolbox loaded with hammers, nails, and other carpentry basics. Musk would visit now and again from Los Angeles and was unfazed by the conditions, having seen SpaceX grow up in similar surroundings.

The original plan for producing a prototype vehicle sounded simple. Tesla would take the AC Propulsion tzero powertrain and fit it into the Lotus Elise body. The company had acquired a schematic for an electric motor design and figured it could buy a transmission from a company in the United States or Europe and outsource any other parts from Asia. Tesla's engineers mostly needed to focus on developing the battery pack systems, wiring the car, and cutting and welding metal as needed to bring everything together. Engineers love to muck around with hardware, and the Tesla team thought of the Roadster as something akin to a car conversion project that could be done with two or three mechanical engineers, and a few assembly people.

The main team of prototype builders consisted of Straubel, Berdichevsky, and David Lyons, a very clever mechanical engineer and employee No. 12. Lyons had about a decade of experience working for Silicon Valley companies and had met Straubel a few years before when the two men struck up a conversation at a 7-Eleven about an electric bike Straubel was riding. Lyons had helped Straubel pay bills by hiring him as a consultant for a company building a device to measure people's core body temperature. Straubel thought he could return the favor by bringing Lyons on early to such an exciting project. Tesla would benefit in a big way as well. As Berdichevsky put it, "Dave Lyons knew how to get shit done."

The engineers bought a blue lift for the car and set it up inside the building. They also purchased some machine tools,

hand tools, and floodlights to work at night and started to turn
the facility into a hotbed of R&D activity. Electrical engineers
studied the Lotus's base-level software to figure out how it tied
together the pedals, mechanical apparatus, and the dashboard
gauges. The really advanced work took place with the battery
pack design. No one had ever tried to combine hundreds of lithium ion batteries in parallel, so Tesla ended up at the cutting edge
of the technology.

The engineers started trying to understand how heat would
dissipate and current flow would behave across seventy batteries
by supergluing them together into groups called bricks. Then ten
bricks would be placed together, and the engineers would test
various types of air and liquid cooling mechanisms. When the
Tesla team had developed a workable battery pack, they stretched
the yellow Lotus Elise chassis five inches and lowered the pack
with a crane into the back of the car, where its engine would normally be. These efforts began in earnest on October 18, 2004,
and, rather remarkably, four months later, on January 27, 2005,
an entirely new kind of car had been built by eighteen people. It
could even be driven around. Tesla had a board meeting that day,
and Musk zipped about in the car. He came away happy enough
to keep investing. Musk put in $9 million more as Tesla raised a
$13 million funding round. The company now planned to deliver
the Roadster to consumers in early 2006.

Once they'd finished building a second car a few months later,
the engineers at Tesla decided they needed to face up to a massive
potential flaw in their electric vehicle. On July 4, 2005, they were
at Eberhard's house in Woodside celebrating Independence Day
and figured it was as good a moment as any to see what happened
when the Roadster's batteries caught on fire. Someone taped
twenty of the batteries together, put a heating strip wire into the
bundle, and set it off. "It went up like a cluster of bottle rock-

ets," Lyons said. Instead of twenty batteries, the Roadster would have close to 7,000, and the thought of what an explosion at that scale would be like horrified the engineers. One of the perks of an electric car was meant to be that it moved people away from a flammable liquid like gasoline and the endless explosions that take place in an engine. Rich people were unlikely to pay a high price for something even more dangerous, and the early nightmare scenario for the employees at Tesla was that a rich, famous person would get caught in a fire caused by the car. "It was one of those 'oh shit' moments," Lyons said. "That is when we really sobered up."

Tesla formed a six-person task force to deal with the battery issue. They were pulled off all other work and given money to begin running experiments. The first explosions started taking place at the Tesla headquarters, where the engineers filmed them in slow motion. Once saner minds prevailed, Tesla moved its explosion research to a blast area behind an electrical substation maintained by the fire department. Blast by blast, the engineers learned a great deal about the inner workings of the batteries. They developed methods for arranging them in ways that would prevent fires spreading from one battery to the next and other techniques for stopping explosions altogether. Thousands of batteries exploded along the way, and the effort was worth it. It was still early days, for sure, but Tesla was on the verge of inventing battery technology that would set it apart from rivals for years to come and would become one of the company's great advantages.

The early success at building two prototype cars, coupled with Tesla's engineering breakthroughs around the batteries and other technological pieces, boosted the company's confidence. It was time to put Tesla's stamp on the vehicle. "The original plan had been to do the bare minimum we could get away with as far as making the car stylistically different from a Lotus but electric," said Tarpenning. "Along the way, Elon and the rest of the board said, 'You only get to do this once. It has to delight the customer, and the Lotus just isn't good enough to do that.' "

The Elise's chassis, or base frame, worked fine for Tesla's engineering purposes. But the body of the car had serious issues in both form and function. The door on the Elise was all of a foot tall, and you were meant to either jump into the car or fall into it, depending on your flexibility and/or dignity. The body also needed to be longer to accommodate Tesla's battery pack and a trunk. And Tesla preferred to make the Roadster out of carbon fiber instead of fiberglass. On these design points, Musk had a lot of opinion and influence. He wanted a car that Justine could feel comfortable getting into and that had some measure of practicality. Musk made these opinions clear when he visited Tesla for board meetings and design reviews.

Tesla hired a handful of designers to mock up new looks for the Roadster. After settling on a favorite, the company paid to build a quarter-scale model of the vehicle in January 2005 and then a full-scale model in April. This process provided the Tesla executives with yet another revelation of everything that went into making a car. "They wrap this shiny Mylar material around the model and vacuum it, so that you can really see the contours and shine and shadows," Tarpenning said. The silver model was then turned into a digital rendering that the engineers could manipulate on their computers. A British company took the digital file and used it to create a plastic version of the car called an "aero buck" for aerodynamics testing. "They put it on a boat and shipped it to us, and then we took it to Burning Man," Tarpenning said, referring to the annual drug-infused art festival held in the Nevada desert.

About a year later, after many tweaks and much work, Tesla had a pencils-down moment. It was May 2006, and the company had grown to a hundred employees. This team built a black version of the Roadster known as EP1, or engineering prototype one. "It was saying, 'We now think we know what we will build,'" Tarpenning said. "You can feel it. It's a real car, and it's very exciting." The arrival of the EP1 provided a great excuse to show existing investors what their money had bought and to ask for more funds from a wider audience. The venture capitalists were impressed enough to overlook the fact that engineers sometimes had to manually fan the car to cool it down in between test drives and were now starting to grasp Tesla's long-term potential. Musk once again put money into Tesla—$12 million—and a handful of other investors, including the venture capital firm Draper Fisher Jurvetson, VantagePoint Capital Partners, J.P. Morgan, Compass Technology Partners, Nick Pritzker, Larry Page, and Sergey Brin, joined the $40 million round.[*]

In July 2006, Tesla decided to tell the world what it had been up to. The company's engineers had built a red prototype—EP2—to complement the black one, and they both went on display at an event in Santa Clara. The press flocked to the announcement and were quite taken with what they saw. The Roadsters were gorgeous, two-seater convertibles that could go from zero to 60 in about four seconds. "Until today," Musk said at the event, "all electric cars have sucked."[6]

Celebrities like then-governor Arnold Schwarzenegger and former Disney CEO Michael Eisner showed up at the event, and many of them took test rides in the Roadsters. The vehicles were

---

[*]   In a press release announcing the funding round, Musk was not listed as a founder of the company. In the "About Tesla Motors" section, the company stated, "Tesla Motors was founded in June 2003 by Martin Eberhard and Marc Tarpenning to create efficient electric cars for people who love to drive." Musk and Eberhard would later spar over Musk's founder status.

so fragile that only Straubel and a couple of other trusted hands knew how to run them, and they were swapped out every five minutes to avoid overheating. Tesla revealed that each car would cost about $90,000 and had a range of 250 miles per charge. Thirty people, the company said, had committed to buying a Roadster, including the Google cofounders Brin and Page and a handful of other technology billionaires. Musk promised that a cheaper car—a four-seat, four-door model under $50,000, would arrive in about three years.

Around the time of this event, Tesla made its debut in the *New York Times* via a mini-profile on the company. Eberhard vowed—optimistically—to begin shipments of the Roadster in the middle of 2007, instead of early 2006 as once planned, and laid out Tesla's strategy of starting with a high-priced, low-volume product and moving down to more affordable products over time, as underlying technology and manufacturing capabilities advanced. Musk and Eberhard were big believers in this strategy, having seen it play out with a number of electronic devices. "Cellphones, refrigerators, color TV's, they didn't start off by making a low-end product for masses," Eberhard told the paper.[7] "They were relatively expensive, for people who could afford it." While the story was a coup for Tesla, Musk didn't appreciate being left out of the article entirely. "We tried to emphasize him, and told the reporter about him over and over again, but they weren't interested in the board of the company," Tarpenning said. "Elon was furious. He was livid."

You could understand why Musk might want some of the shine of Tesla to rub off on him. The car had turned into a cause célèbre of the automotive world. Electric vehicles tended to invoke religious overreactions from both the pro and con camps, and the appearance of a good-looking, fast electric car stoked everyone's passions. Tesla had also turned Silicon Valley into a real threat, at least conceptually, to Detroit for the first time. The month after the Santa Monica event was the Pebble Beach Concours d'Elegance, a famous showcase for exotic cars. Tesla had become such a topic of conversation that the organizers of the event begged to have a Roadster and waived the usual display fees. Tesla set up a booth, and people showed up by the dozens writing $100,000 checks on the spot to pre-order their cars. "This was long before Kickstarter, and we just had not thought of trying to do that," Tarpenning said. "But then we started getting millions of dollars at these types of events." Venture capitalists, celebrities, and friends of Tesla employees began trying to buy their way onto the waiting list. Some of Silicon Valley's wealthy elite went so far as to show up at the Tesla office and knock on the door, looking to buy a car. The entrepreneurs Konstantin Othmer and Bruce Leak, who had known Musk from his internship days at Rocket Science Games, did just that one weekday and ended up getting a personal tour of the car from Musk and Eberhard that stretched over a couple of hours. "At the end we said, 'We'll take one,'" Othmer said. "They weren't actually allowed to sell cars yet, though, so we joined their club. It cost one hundred thousand dollars, but one of the benefits of membership was that you'd get a free car."

As Tesla switched from marketing back into R&D mode, it had some trends working in its favor. Advances in computing had made it so that small car companies could sometimes punch at the same weight as the giants of the industry. Years ago, automakers would have needed to make a fleet of cars for crash testing. Tesla could not afford to do that, and it didn't have to. The third Roadster engineering prototype went to the same collision testing facility used by large automakers, giving Tesla access to top-of-the-line high-speed cameras and other imaging technology. Thousands of other tests, though, were done by a third party that specialized in computer simulations and saved Tesla from building a fleet of crash vehicles. Tesla also had equal access to the big guys' durability tracks made out of cobblestones and concrete embedded with metal objects. It could replicate 100,000 miles and ten years of wear at these facilities.

Quite often, the Tesla engineers brought their Silicon Valley attitude to the automakers' traditional stomping grounds. There's a break and traction testing track in northern Sweden near the Arctic Circle where cars get tuned on large plains of ice. It would be standard to run the car for three days or so, get the data, and return to company headquarters for many weeks of meetings about how to adjust the car. The whole process of tuning a car can take the entire winter. Tesla, by contrast, sent its engineers along with the Roadsters being tested and had them analyze the data on the spot. When something needed to be tweaked, the engineers would rewrite some code and send the car back on the ice. "BMW would need to have a confab between three or four companies that would all blame each other for the problem," Tarpenning said. "We just fixed it ourselves." Another testing procedure required that the Roadsters go into a special cooling chamber to check how they would respond to frigid temperatures. Not wanting to pay the exorbitant costs to use one of these chambers, the Tesla engineers opted to rent an ice cream delivery truck with a large refrigerated trailer. Someone would drive a Roadster into the truck, and the engineers would don parkas and work on the car.

Every time Tesla interacted with Detroit it received a reminder of how the once-great city had been separated from its own can-do culture. Tesla tried to lease a small office in Detroit. The costs were incredibly low compared with space in Silicon Valley, but the city's bureaucracy made getting just a basic office an ordeal. The building's owner wanted to see seven years of audited financials from Tesla, which was still a private company. Then the building owner wanted two years' worth of advanced rent. Tesla had about $50 million in the bank and could have bought the building outright. "In Silicon Valley, you say you're backed by a venture capitalist, and that's the end of the negotiation," Tarpenning said. "But everything was like that in Detroit. We'd get FedEx boxes, and they couldn't even decide who should sign for the package."

Throughout these early years, the engineers credited Eberhard with making quick, crisp decisions. Rarely did Tesla get hung up overanalyzing a situation. The company would pick a plan of attack, and when it failed at something, it failed fast and then tried a new approach. It was many of the changes that Musk wanted that started to delay the Roadster. Musk kept pushing for the car to be more comfortable, asking for alterations to the seats and the doors. He made the carbon-fiber body a priority, and he pushed for electronic sensors on the doors so that the Roadster could be unlocked with the touch of a finger instead of a tug on a handle. Eberhard groused that these features were slowing the company down, and many of the engineers agreed. "It felt at times like Elon was this unreasonably demanding overarching force," said Berdichevsky. "The company as a whole was sympathetic to Martin because he was there all the time, and we all felt the car should ship sooner."

By the middle of 2007, Tesla had grown to 260 employees and seemed to be pulling off the impossible. It had produced the fastest, most beautiful electric car the world had ever seen almost from thin air. All it had to do next was build a lot of the cars—a process that would end up almost bankrupting the company.

The greatest mistake Tesla's executives made in the early days were assumptions around the transmission system for the Roadster. The goal had always been to get from zero to 60 mph as quickly as possible in the hopes that the raw speed of the Roadster would attract a lot of attention and make it fun to drive. To do this, Tesla's engineers had decided on a two-speed transmission, which is the underlying mechanism in the car for transferring power from the motor to the wheels. The first gear would take the car from zero to 60 mph in less than four seconds, and then the second gear would take the car up to 130 mph. Tesla had hired a company specializing in transmission designs to build this part and had every reason to believe that this would be one of the smoother bits of the Roadster's journey. "People had been making transmissions since Robert Fulton built the steam engine," said Bill Currie,[8] a veteran Silicon Valley engineer and employee No. 86 at Tesla. "We thought you would just order one. But the first one we had lasted forty seconds." The initial transmission could not handle the big jump from the first to the second gear, and the fear was that the second gear would engage at high speed and not be synchronized with the motor properly, which would result in catastrophic damage to the car.

Lyons and the other engineers quickly set out to try to fix the issue. They found a couple of other contractors to design replacements and again hoped that these longtime transmission experts would deliver something usable with relative ease. It soon became apparent, however, that the contractors were not always putting their A team to work on this project for a tiny start-up in Silicon Valley and that the new transmissions were no better than the first. During tests, Tesla found that the transmissions would sometimes break after 150 miles and that the mean time between failures was about 2,000 miles. When a team from Detroit ran a root cause analysis of the transmission to find failures, they discovered fourteen separate issues that could cause the system to break. Tesla had wanted to deliver the Roadster in November 2007, but the transmission issues lingered, and by the time January 1, 2008, rolled around, the company had to once again start from scratch, on a third transmission push.

Tesla also faced issues abroad. The company had decided to send a team of its youngest, most energetic engineers to Thailand to set up a battery factory. Tesla partnered with an enthusiastic although not totally capable manufacturing partner. The Tesla engineers had been told that they could fly over and manage the construction of a state-of-the-art battery factory. Instead of a factory, they found a concrete slab with posts holding up a roof. The building was about a three-hour drive south from Bangkok, and had been left mostly open like many of the other factories because of the incredible heat. The other manufacturing operations dealt with making stoves, tires, and commodities that could withstand the elements. Tesla had sensitive batteries and electronics, and like parts of the Falcon 1, they'd be chewed up by the salty, humid conditions. Eventually, Tesla's partner paid about $75,000 to put in drywall, coat the floor, and create storage rooms with temperature controls. Tesla's engineers ended up working maddening hours trying to train the Thai workers on how to handle the electronics properly. The development of the battery technology, which had once moved along at a rapid pace, slowed to a crawl.

The battery factory was one part of a supply chain that stretched across the globe, adding cost and delays to the Roadster production. Body panels for the car were to be made in France, while the motors were to come from Taiwan. Tesla planned on buying battery cells in China and shipping them to Thailand to turn the piece parts into battery packs. The battery packs, which had to be stored for a minimal amount of time to avoid degradation, would then be taken to port and shipped to England, where they needed to clear customs. Tesla then planned for Lotus to build the body of the car, attach the battery packs, and ship the

Roadsters by boat around Cape Horn to Los Angeles. In that scenario, Tesla would have paid for the bulk of the car and had no chance to recognize revenue on the parts until six to nine months had passed. "The idea was to get to Asia, get things done fast and cheap, and make money on the car," said Forrest North, one of the engineers sent to Thailand. "What we found out was that for really complicated things, you can do the work cheaper here and have less delays and less problems." When some new hires came on, they were horrified to discover just how haphazard Tesla's plan appeared. Ryan Popple, who had spent four years in the army and then gotten an MBA from Harvard, arrived at Tesla as a director of finance meant to prep the company to go public. After examining the company's books early in his tenure, Popple asked the manufacturing and operations head exactly how he would get the car made. "He said, 'Well, we will decide we're going into production and then a miracle is going to happen,'" Popple said.

As word of the manufacturing issues reached Musk, he became very concerned about the way Eberhard had run the company and called in a fixer to address the situation. One of Tesla's investors was Valor Equity, a Chicago-based investment firm that specialized in fine-tuning manufacturing operations. The company had been drawn to Tesla's battery and powertrain technology and calculated that even if Tesla failed to sell many cars, the big automakers would end up wanting to buy its intellectual property. To protect its investment, Valor sent in Tim Watkins, its managing director of operations, and he soon reached some horrific conclusions.

Watkins is a Brit with degrees in industrial robotics and electrical engineering. He's built up a reputation as an ingenious solver of problems. While doing work in Switzerland, for example, Watkins found a way to get around the country's rigid labor laws that limit the hours employees can work, by automating a metal stamping factory so that it could run twenty-four hours per day instead of sixteen hours like the factories or rivals. Watkins is also known for keeping his ponytail in place with a black scrunchie, wearing a black leather jacket, and toting a black fanny pack everywhere he goes. The fanny pack has his passport, checkbook, earplugs, sunscreen, food, and an assortment of other necessities. "It's full of the everyday things I need to survive," said Watkins. "If I walk ten feet away from this thing, I sense it." While a bit eccentric, Watkins was thorough and spent weeks talking to employees and analyzing every part of Tesla's supply chain to figure out how much it cost to make the Roadster.

Tesla had done a decent job of keeping its employee costs down. It hired the kid fresh out of Stanford for $45,000 rather than the proven guy who probably didn't want to work that hard anyway for $120,000. But when it came to equipment and materials, Tesla was a spending horror show. No one liked using the company's software that tracked the bill of materials. So some people used it, and some people didn't. Those that did use it often made huge errors. They would take the cost of a part from the prototype cars and then estimate how much of a discount they expected when buying that part in bulk, rather than actually negotiating to find a viable price. At one point, the software declared that each Roadster should cost about $68,000, which would leave Tesla making about $30,000 per vehicle. Everyone knew the figure was wrong, but it got reported to the board anyway.

Around the middle of 2007, Watkins came to Musk with his findings. Musk was prepared for a high figure but felt confident that the price of the car would come down significantly over time as Tesla ironed out its manufacturing process and increased its sales. "That's when Tim told me it was really bad news," Musk said. It looked like each Roadster could cost up to $200,000 to make, and Tesla planned to sell the car for only around $85,000. "Even in full production, they would have been like $170,000 or something insane," Musk said. "Of course, it didn't much matter because about a third of the cars didn't flat-out fucking work."

Eberhard made attempts to pull his team out of this mess. He'd gone to see a speech in which the famous venture capitalist John Doerr, who became a major investor in green technology companies, declared that he would devote his time and money to trying to save the Earth from global warming because he owed such an effort to his children. Eberhard promptly returned to the Tesla building and ginned up a similar speech. In front of about a hundred people, Eberhard had a picture of his young daughter projected onto the wall of the main workshop. He asked the Tesla engineers why he had put that picture up. One of them guessed that it was because people like his daughter would drive the car. To which Eberhard replied, "No. We are building this because by the time she is old enough to drive she will know a car as something completely different to how we know it today, just like you don't think of a phone as a thing on the wall with a cord on it. It's this future that depends on you." Eberhard then thanked some of the key engineers and called out their efforts in public. Many of the engineers had been pulling all-nighters on a regular basis and Eberhard's show boosted morale. "We were all working ourselves to the point of exhaustion," said David Vespremi, a former Tesla spokesman. "Then came this profound moment where we were reminded that building the car was not about getting to an IPO or selling it to a bunch of rich dudes but because it might change what a car is."

These victories, though, were not enough to overcome the feeling shared by many of the Tesla engineers that Eberhard had reached the end of his abilities as a CEO. The company veterans had always admired Eberhard's engineering smarts and continued to do so. Eberhard, in fact, had turned Tesla into a cult of engineering. Regrettably, other parts of the company had been neglected, and people doubted Eberhard's ability to take the company from the R&D stage to production. The ridiculous cost of the car, the transmission, the ineffective suppliers were crippling Tesla. And, as the company started to miss its delivery dates, many of the once-fanatical consumers who had made their large up-front payments turned on Tesla and Eberhard. "We saw the writing on the wall," Lyons said. "Everyone knew that the person who starts a company is not necessarily the right person to lead it in the long term, but whenever that is the case, it's not easy."

Eberhard and Musk had battled for years over some of the design points on the car. But for the most part, they had gotten along well enough. Neither man suffered fools. And they certainly shared many of the same visions for the battery technology and what it could mean to the world. What their relationship could not survive were the cost figures for the Roadster unearthed by Watkins. It looked to Musk as if Eberhard had grossly mismanaged the company by allowing the parts costs to soar so high. Then, as Musk saw it, Eberhard failed to disclose the severity of the situation to the board. While on his way to give a talk to the Motor Press Guild in Los Angeles, Eberhard received a call from Musk and in a brief, uncomfortable chat learned that he would be replaced as CEO.

In August 2007, Tesla's board demoted Eberhard and named him president of technology, which only exacerbated the company's issues. "Martin was so bitter and disruptive," Straubel said. "I remember him running around the office and sowing discontent, as we're trying to finish the car and are running out of money and everything is at knife's edge." As Eberhard saw it, other people at Tesla had foisted a wonky finance software application on him that made it tricky to accurately track costs. He contended that the delays and cost increases were partly due to the requests of other members of the management team and that he'd been up front with the board about the issues. Beyond that, he thought Watkins had made the situation out to be worse than it really was. Start-ups in Silicon Valley view mayhem as standard operating procedure. "Valor was used to dealing with older companies," Eberhard said. "They found chaos and weren't used to it. This was the chaos of a start-up." Eberhard had also already been asking Tesla's board to replace him as CEO and find someone with more manufacturing experience.

A few months passed, and Eberhard remained pissed-off. Many of the Tesla employees felt like they were caught in the middle of a divorce and had to pick their parent—Eberhard or Musk. By the time December arrived, the situation was untenable, and Eberhard left the company altogether. Tesla said in a statement that Eberhard had been offered a position on its advisory board, although he denied that. "I am no longer with Tesla Motors—neither on its board of directors nor an employee of any sort," Eberhard said in a statement at the time. "I'm not happy with the way I was treated." Musk sent a note to a Silicon Valley newspaper saying, "I'm sorry that it came to this and wish it were not so. It was not a question of personality differences, as the decision to have Martin transition to an advisory role was unanimous among the board. Tesla has operational problems that need to be solved and if the board thought there was any way that Martin could be part of the solution, then he would still be an employee of the company."[9] These statements were the start of a war that would drag on between the two men in public for years and that in many ways continues to the present day.

As 2007 played out, the problems mounted for Tesla. The carbon-fiber body that looked so good turned out to be a huge pain to paint, and Tesla had to cycle through a couple of companies to find one that could do the work well. Sometimes there were faults in the battery pack. The motor short-circuited now and again. The body panels had visible gaps. The company also had to face up to the reality that a two-speed transmission was not going to happen. In order for the Roadster to achieve its flashy zero-to-60 times with a single-speed transmission, Tesla's engineers had to redesign the car's motor and inverter and shave off some weight. "We essentially had to do a complete reboot," Musk said. "That was terrible."

After Eberhard was removed as CEO, Tesla's board tapped Michael Marks as its interim chief. Marks had run Flextronics, an enormous electronics supplier, and had deep experience with complex manufacturing operations and logistics issues. Marks began interrogating various groups at the company to try to figure out their problems and to prioritize the issues plaguing the Roadster. He also put in some basic rules like making sure that people all showed up at work at the same time to establish a baseline of productivity—a tricky ask in Silicon Valley's work anywhere, anytime culture. All of these moves were part of the Marks List, a 10-point, 100-day plan that included eliminating all faults in the battery packs, getting gaps between body parts to less than 40 mm, and booking a specified number of reservations. "Martin had been falling apart and lacked a lot of the discipline key for a manager," Straubel said. "Michael came in and evaluated the mess and was a bullshit filter. He didn't really have a dog in the fight and could say, 'I don't care what you think or what you think. This is what we should do.'" For a while, Marks's strategy worked, and the engineers at Tesla could once again focus on building the Roadster rather than on internal politics. But then Marks's vision for the company began to diverge from Musk's.

By this time, Tesla had moved into a larger facility at 1050 Bing Street in San Carlos. The bigger building allowed Tesla to bring the battery work back in-house from Asia and for it to do some of the Roadster manufacturing, alleviating the supply chain issues. Tesla was maturing as a car company, although its wild-child start-up streak remained well intact. While strolling around the factory one day, Marks saw a Smart car from Daimler on a lift. Musk and Straubel had a small side project going on around the Smart car to see what it might be like as an electric vehicle. "Michael didn't know about it, and he's like, 'Who is the CEO here?'" said Lyons. (The work on the Smart car eventually led to Daimler buying a 10 percent stake in Tesla.)

Marks's inclination was to try to package Tesla as an asset that could be sold to a larger car company. It was a perfectly reasonable plan. While running Flextronics, Marks had overseen a vast, global supply chain and knew the difficulties of manufacturing intimately. Tesla must have looked borderline hopeless to him at this point. The company could not make its one product well, was poised to hemorrhage money, and had missed a string of delivery deadlines and yet its engineers were still off doing side experiments. Making Tesla look as pretty as possible for a suitor was the rational thing to do.

In just about every other case, Marks would be thanked for his decisive plan of action and saving the company's investors from a big loss. But Musk had little interest in polishing up Tesla's assets for the highest bidder. He'd started the company to put a dent in the automotive industry and force people to rethink electric cars. Instead of doing the fashionable Silicon Valley thing of "pivoting" toward a new idea or plan, Musk would dig in deeper. "The product was late and over budget and everything was wrong, but Elon didn't want anything to do with those plans to either sell the whole company or lose control through a partnership," Straubel said. "So, Elon decided to double down."

On December 3, 2007, Ze'ev Drori replaced Marks as CEO.

Drori had experience in Silicon Valley starting a company that made computer memory and selling it to the chipmaker Advanced Micro Devices. Drori was not Musk's first pick—a top choice had turned down the job because he didn't want to move from the East Coast—and did not inspire much enthusiasm from the Tesla employees. Drori had about fifteen years on the youngest Tesla worker and no connection to this group bonded by suffering and toil. He came to be seen more as an executor of Musk's wishes than as a commanding, independent CEO.

Musk began making more public gestures to mitigate the bad press around Tesla. He issued statements and did interviews, promising that the Roadster would ship to customers in early 2008. He began talking up a car code-named WhiteStar—the Roadster had been code-named DarkStar—that would be a sedan possibly priced around $50,000, and a new factory to build the machine. "Given the recent management changes, some reassurances are in order regarding Tesla Motors' future plans," Musk wrote in a blog post. "The near term message is simple and unequivocal—we are going to deliver a great sports car next year that customers will love driving. . . . My car, production VIN 1, is already off the production line in the UK and final preparations are being made for importation." Tesla held a series of town hall meetings with customers where it tried to fess up to its problems in the open, and it started building some showrooms for its car. Vince Sollitto, the former PayPal executive, visited the Menlo Park showroom and found Musk complaining about the public relations issues but clearly inspired by the product Tesla was building. "His demeanor changed the moment we got to this display of the motor," Sollitto said. Dressed in a leather jacket and slacks, Musk started talking about the motor's properties and then put on a performance worthy of a carnival strongman by lifting the hundred-or-so-pound hunk of metal. "He picks this thing up and wedges it between his two palms," Sollitto said. "He's holding it, and he's shaking and beads of sweat are forming on his forehead. It wasn't so much a display of strength as a physical demonstration of the beauty of the product." While the customers complained a lot about the delays, they seemed to sense this passion from Musk and share his enthusiasm for the product. Only a handful of customers asked for their prepayments back.

Tesla employees soon got to witness the same Musk that SpaceX employees had seen for years. When an issue like the Roadster's faulty carbon-fiber body panels cropped up, Musk dealt with it directly. He flew to England in his jet to pick up some new manufacturing tools for the body panels and personally delivered them to a factory in France to ensure that the Roadster stayed on its production schedule. The days of people being ambiguous about the Roadster's manufacturing costs were gone as well. "Elon got fired up and said we were going to do this intense cost-down program," said Popple. "He gave a speech, saying we would work on Saturdays and Sundays and sleep under desks until it got done. Someone pushed back from the table and argued that everyone had been working so hard just to get the car done, and they were ready for a break and to see their families. Elon said, 'I would tell those people they will get to see their families a lot when we go bankrupt.' I was like, 'Wow,' but I got it. I had come out of a military culture, and you just have to make your objective happen." Employees were required to meet at 7 A.M. every Thursday morning for bill-of-materials updates. They had to know the price of every part and have a cogent plan for getting parts cheaper. If the motor cost $6,500 a pop at the end of December, Musk wanted it to cost $3,800 by April. The costs were plotted and analyzed each month. "If you started falling behind, there was hell to pay," Popple said. "Everyone could see it, and people lost their jobs when they didn't deliver. Elon has a mind that's a bit like a calculator. If you put a number on the projector that does not make sense, he will spot it. He doesn't miss details." Popple found Musk's style aggressive, but he liked that Musk would listen to a well-argued, analytical point and often change his mind if given a good enough reason. "Some people thought Elon was too tough or hot-tempered or tyrannical," Popple said. "But these were hard times, and those of us close to the operational realities of the company knew it. I appreciated that he didn't sugarcoat things."

On the marketing front, Musk would run daily Google searches for news stories about Tesla. If he saw a bad story, he ordered someone to "fix it" even though the Tesla public relations people could do little to sway the reporters. One employee missed an event to witness the birth of his child. Musk fired off an e-mail saying, "That is no excuse. I am extremely disappointed. You need to figure out where your priorities are. We're changing the world and changing history, and you either commit or you don't."*

Marketing people who made grammatical mistakes in e-mails were let go, as were other people who hadn't done anything "awesome" in recent memory. "He can be incredibly intimidating at times but doesn't have a real sense for just how imposing he can be," said one former Tesla executive. "We'd have these meetings and take bets on who was going to get bloodied and bruised. If you told him that you made a particular choice because 'it was the standard way things had always been done,' he'd kick you out of a meeting fast. He'd say, 'I never want to hear that phrase again. What we have to do is fucking hard and half-assing things won't

---

* This was how the employee remembered the text. I did not see the actual e-mail. Musk later told the same employee, "I want you to think ahead and think so hard every day that your head hurts. I want your head to hurt every night when you go to bed."

be tolerated.' He just destroys you and, if you survive, he determines if he can trust you. He has to understand that you're as crazy as he is." This ethos filtered through the entire company, and everyone quickly understood that Musk meant business.

Straubel, while sometimes on the bad end of the critiques, welcomed Musk's hard-charging presence. The five years to get to this point had been an enjoyable slog for him. Straubel had transformed from a quiet, capable engineer who shuffled around Tesla's factory floor with his head down into the most crucial member of the technical team. He knew more about the batteries and the electric drivetrain than just about anyone else at the company. He also began developing a role as a go-between for employees and Musk. Straubel's engineering smarts and work ethic had earned Musk's respect, and Straubel found that he could deliver difficult messages to Musk on behalf of other employees. As he would do for years to come, Straubel also proved willing to check his ego at the door. All that mattered was getting the Roadster and the follow-on sedan to market to popularize electric cars, and Musk looked like the best person to make that happen.

Other employees had enjoyed the thrill of the engineering challenge over the past five years but were burnt-out beyond repair. Wright didn't believe that an electric car for the masses would ever take off. He left and started his own company dedicated to making electric versions of delivery trucks. Berdichevsky had been a crucial, do-anything young engineer for much of Tesla's existence. Now that the company employed about three hundred people, he felt less effective and didn't relish the idea of suffering for another five years to bring the sedan to market. He would leave Tesla, get a couple of degrees from Stanford, and cofound a start-up looking to make a revolutionary new battery that could soon go into electric cars. With Eberhard gone, Tarpenning found Tesla less fun. He didn't see eye to eye with Drori and also shied away from the idea of frying his soul to get the sedan out. Lyons stuck around longer, which is a minor miracle. At various points, he had led the development of most of the core technology behind the Roadster, including the battery packs, the motor, the power electronics, and, yes, the transmission. This meant that for about five years Lyons had been among Tesla's most capable employees and the guy constantly in the doghouse for being behind on something and thus holding the rest of the company up. He suffered through some of Musk's more colorful tirades—directed either at him or suppliers that had let Tesla down—that included talk of people's balls being chopped off and other violent or sexual acts. Lyons also saw an exhausted, stressed-out Musk spit coffee across a conference room table because it was cold and then, without a pause, demand that the employees work harder, do more, and mess up less. Like so many people privy to these performances, Lyons came away with no illusions about Musk's personality but with the utmost respect for his vision and drive to execute. "Working at Tesla back then was like being Kurtz in *Apocalypse Now*," Lyons said. "Don't worry about the methods or if they're unsound. Just get the job done. It comes from Elon. He listens, asks good questions, is fast on his feet, and gets to the bottom of things."

Tesla could survive the loss of some of these early hires. Its strong brand had allowed the company to keep recruiting top talent, including people from large automotive companies who knew how to get over the last set of challenges blocking the Roadster from reaching customers. But Tesla's major issue no longer revolved around effort, engineering, or clever marketing. Heading into 2008, the company was running out of money. The Roadster had cost about $140 million to develop, way over the $25 million originally estimated in the 2004 business plan.

8

PAIN, SUFFERING, AND SURVIVAL

AS HE PREPARED TO BEGIN FILMING *IRON MAN* IN EARLY 2007, THE DIREC-
tor Jon Favreau rented out a complex in Los Angeles that
once belonged to Hughes Aircraft, the aerospace and defense
contractor started about eighty years earlier by Howard Hughes.
The facility had a series of interlocking hangars and served as a
production office for the movie. It also supplied Robert Downey
Jr., who was to play *Iron Man* and his human creator Tony Stark,
with a splash of inspiration. Downey felt nostalgic looking at one
of the larger hangars, which had fallen into a state of disrepair.
Not too long ago, that building had played host to the big ideas
of a big man who shook up industries and did things his own way.

Downey heard some rumblings about a Hughes-like figure
named Elon Musk who had constructed his own, modern-day
industrial complex about ten miles away. Instead of visualizing
how life might have been for Hughes, Downey could perhaps get
a taste of the real thing. He set off in March 2007 for SpaceX's
headquarters in El Segundo and wound up receiving a personal

tour from Musk. "My mind is not easily blown, but this place and this guy were amazing," Downey said.

To Downey, the SpaceX facility looked like a giant, exotic hardware store. Enthusiastic employees were zipping about, fiddling with an assortment of machines. Young white-collar engineers interacted with blue-collar assembly line workers, and they all seemed to share a genuine excitement for what they were doing. "It felt like a radical start-up company," Downey said. After the initial tour, Downey came away pleased that the sets being hammered out at the Hughes factory did have parallels to the SpaceX factory. "Things didn't feel out of place," he said.

Beyond the surroundings, Downey really wanted a peek inside Musk's psyche. The men walked, sat in Musk's office, and had lunch. Downey appreciated that Musk was not a foul-smelling, fidgety, coder whack job. What Downey picked up on instead were Musk's "accessible eccentricities" and the feeling that he was an unpretentious sort who could work alongside the people in the factory. Both Musk and Stark were the type of men, according to Downey, who "had seized an idea to live by and something to dedicate themselves to" and were not going to waste a moment.

When he returned to the *Iron Man* production office, Downey asked that Favreau be sure to place a Tesla Roadster in Tony Stark's workshop. On a superficial level, this would symbolize that Stark was so cool and connected that he could get a Roadster before it even went on sale. On a deeper level, the car was to be placed as the nearest object to Stark's desk so that it formed something of a bond between the actor, the character, and Musk. "After meeting Elon and making him real to me, I felt like having his presence in the workshop," Downey said. "They became contemporaries. Elon was someone Tony probably hung out with and partied with or more likely they went on some weird jungle trek together to drink concoctions with the shamans."

After *Iron Man* came out, Favreau began talking up Musk's role as the inspiration for Downey's interpretation of Tony Stark. It was a stretch on many levels. Musk is not exactly the type of guy who downs scotch in the back of a Humvee while part of a military convoy in Afghanistan. But the press lapped up the comparison, and Musk started to become more of a public figure. People who sort of knew him as "that PayPal guy" began to think of him as the rich, eccentric businessman behind SpaceX and Tesla.

Musk enjoyed his rising profile. It fed his ego and provided some fun. He and Justine bought a house in Bel Air. Their neighbor to one side was Quincy Jones, the music producer, and their other neighbor was Joe Francis, the infamous creator of the *Girls Gone Wild* videos. Musk and some former PayPal executives, having settled their differences, produced *Thank You for Smoking* and used Musk's jet in the movie. While not a hard-drinking carouser, Musk took part in the Hollywood nightlife and its social scene. "There were just a lot of parties to go to," said Bill Lee, Musk's close friend. "Elon was neighbors with two quasi-celebrities. Our friends were making movies and through this confluence of our networks, there was something to go out and do every night." In one interview, Musk calculated that his life had become 10 percent playboy and 90 percent engineer.[10] "We had a domestic staff of five; during the day our home transformed into a workplace," Justine wrote in magazine article. "We went to black-tie fundraisers and got the best tables at elite Hollywood nightclubs, with Paris Hilton and Leonardo DiCaprio partying next to us. When Google cofounder Larry Page got married on Richard Branson's private Caribbean island, we were there, hanging out in a villa with John Cusack and watching Bono pose with swarms of adoring women outside the reception tent."

Justine appeared to relish their status even more than Musk.

A writer of fantasy fiction novels, she kept a blog detailing the couple's family life and their adventures on the town. In one entry, Justine had Musk saying that he'd prefer to sleep with Veronica than Betty from the *Archie* comics and that he'd like to visit a Chuck E. Cheese sometime. In another entry, she wrote about meeting Leonardo DiCaprio at a club and having him beg for a free Tesla Roadster, only to be turned down. Justine handed out nicknames to oft-occurring characters in the blog, so Bill Lee became "Bill the Hotel Guy" because he owns a hotel in the Dominican Republic, and Joe Francis appeared as "Notorious Neighbor." It's hard to imagine Musk, who keeps to himself, hanging out with someone as ostentatious as Francis, but the men got along well. When Francis took over an amusement park for his birthday, Musk attended and then ended up partying at Francis's house. Justine wrote, "E was there for a bit but admitted he also found it 'kind of lame'—he's been to a couple of parties at NN's house now and ends up feeling self-conscious, 'because it just seems like there are always these skeevy guys wandering around the house trolling for girls. I don't want to be seen as one of those guys.'" When Francis got ready to buy a Roadster, he stopped by the Musks' house and handed over a yellow envelope with $100,000 in cash.

For a while, the blog provided a rare, welcome glimpse into the life of an unconventional CEO. Musk seemed charming. The public learned that he bought Justine a nineteenth-century edition of *Pride and Prejudice*, that Musk's best friends gave him the nickname "Elonius," and that Musk likes to place one-dollar wagers on all manner of things—Can you catch herpes from the Great Barrier Reef? Is it possible to balance two forks with a toothpick?—that he knows he will win. Justine told one story about Musk traveling to Necker Island, in the British Virgin Islands, to hang out with Tony Blair and Richard Branson. A

The Haldeman children had lots of downtime in the African bush while on wild adventures with their parents. ©*Maye Musk*

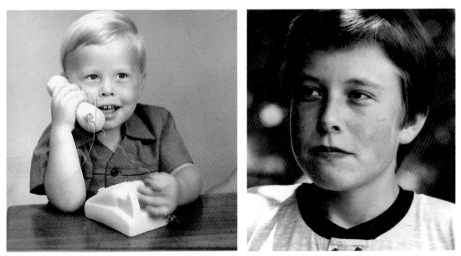

Left: As a toddler, Musk would often drift off into his own world and ignore those around him. Doctors theorized that he might be hard of hearing and had his adenoid glands removed. *Right:* Musk was a loner throughout grade school and suffered for years at the hands of bullies. ©*Maye Musk*

Musk's original video-game code for Blastar, the game he wrote as a twelve-year-old and published in a local magazine. ©*Maye Musk*

(*From left to right:*) Elon, Kimbal, and Tosca at their house in South Africa. All three children now live in the United States. ©*Maye Musk*

Musk ran away on his own to Canada and ended up at Queen's University in Ontario, living in a dormitory for foreign students. ©*Maye Musk*

J. B. Straubel puts together one of Tesla Motors' early battery packs at his house. *Photograph courtesy of Tesla Motors*

A handful of engineers built the first Tesla Roadster in a Silicon Valley warehouse that they had turned into a garage workshop and research lab. *Photograph courtesy of Tesla Motors*

Musk and Martin Eberhard prepare to take the early Roadster for a test-drive. The relationship between the two men would fall apart in the years to come. *Photograph courtesy of Tesla Motors*

SpaceX built its rocket factory from the ground up in a Los Angeles warehouse to give birth to the Falcon 1 rocket. *Photograph courtesy of SpaceX*

Tom Mueller (*far right, gray shirt*) led the design, testing, and construction of SpaceX's engines. *Photograph courtesy of SpaceX*

SpaceX had to conduct its first flights from Kwajalein Atoll (or Kwaj) in the Marshall Islands. The island experience was a difficult but ultimately fruitful adventure for the engineers. *Photograph courtesy of SpaceX*

SpaceX built a mobile mission-control trailer, and Musk and Mueller used it to monitor the later launches from Kwaj. *Photograph courtesy of SpaceX*

Musk hired Franz von Holzhausen in 2008 to design the Tesla Model S. The two men speak almost every day, as can be seen in this meeting in Musk's SpaceX cubicle. ©*Steve Jurvetson*

SpaceX's ambitions grew over the years to include the construction of the Dragon capsule, which could take people to the International Space Station and beyond. ©*Steve Jurvetson*

Musk has long had a thing for robots and is always evaluating new machines for both the SpaceX and Tesla factories. ©*Steve Jurvetson*

When SpaceX moved to a new factory in Hawthorne, California, it was able to scale out its assembly line and work on multiple rockets and capsules at the same time. ©*Steve Jurvetson*

SpaceX tests new engines and crafts at a site in McGregor, Texas. Here the company is testing a reusable rocket, code-named "Grasshopper," that can land itself. *Photograph courtesy of SpaceX*

Musk has a tradition of visiting Dairy Queen ahead of test flights in Texas, in this case with SpaceX investor and board member Steve Jurvetson (*left*) and fellow investor Randy Glein (*right*). ©*Steve Jurvetson*

With a Dragon capsule hanging overhead, SpaceX employees peer into the company's mission control center at the Hawthorne factory. *Photograph courtesy of SpaceX*

Gwynne Shotwell is Musk's right-hand woman at SpaceX and oversees the day-to-day operations of the company, including monitoring a launch from mission control. *Photograph courtesy of SpaceX*

Tesla took over the New United Motor Manufacturing Inc. (or NUMMI) car factory in Fremont, California, which is where workers produce the Model S sedan. *Photograph courtesy of Tesla Motors*

Tesla began shipping the Model S sedan in 2012. The car ended up winning most of the automotive industry's major awards. *Photograph courtesy of Tesla Motors*

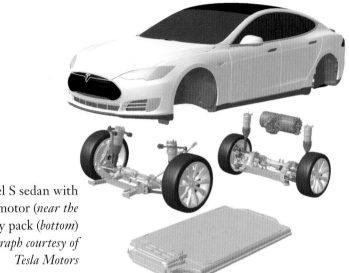

The Tesla Model S sedan with its electric motor (*near the rear*) and battery pack (*bottom*) exposed. *Photograph courtesy of Tesla Motors*

Tesla's next car will be the Model X SUV with its signature "falcon-wing doors." *Photograph courtesy of Tesla Motors*

In 2013, Musk visited Cuba with Sean Penn (*driving*) and the investor Shervin Pishevar (*back seat next to Musk*). They met with students and members of the Castro family, and tried to free an American prisoner. ©*Shervin Pishevar*

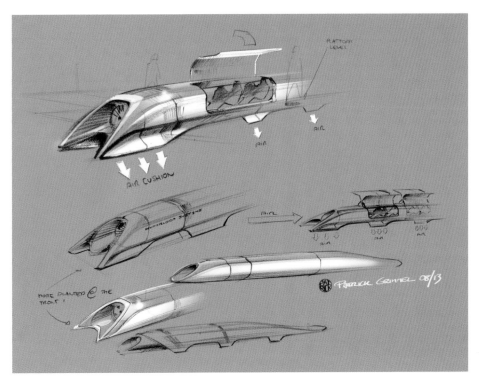

Musk unveiled the Hyperloop in 2013. He proposed it as a new mode of transportation, and multiple groups have now set to work on building it.
Photograph courtesy of SpaceX

In 2014, Musk unveiled a radical new take on the space capsule—the Dragon V2. It comes with a drop-down touch-screen display and slick interior. *Photograph courtesy of SpaceX*

The Dragon V2 will be able to return to Earth and land with pinpoint accuracy. *Photograph courtesy of SpaceX*

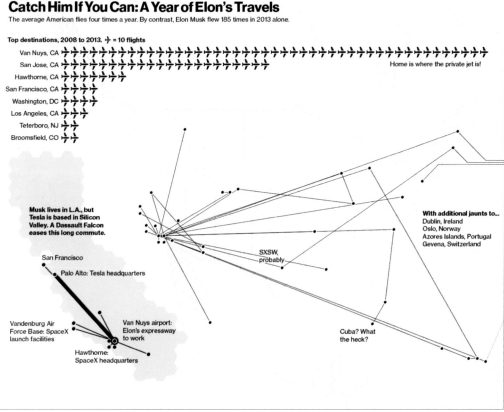

Catch Him If You Can: A Year of Elon's Travels

The average American flies four times a year. By contrast, Elon Musk flew 185 times in 2013 alone.

Top destinations, 2008 to 2013. ✈ = 10 flights

Van Nuys, CA
San Jose, CA
Hawthorne, CA
San Francisco, CA
Washington, DC
Los Angeles, CA
Teterboro, NJ
Broomsfield, CO

Home is where the private jet is!

Musk lives in L.A., but
Tesla is based in Silicon
Valley. A Dassault Falcon
eases this long commute.

With additional jaunts to...
Dublin, Ireland
Oslo, Norway
Azores Islands, Portugal
Gevena, Switzerland

San Francisco

Palo Alto: Tesla headquarters

SXSW,
probably

Vandenburg Air
Force Base: SpaceX
launch facilities

Van Nuys airport:
Elon's expressway
to work

Cuba? What
the heck?

Hawthorne:
SpaceX headquarters

Musk is a nonstop traveler. Here's a look at one year in his life via records
obtained through a Freedom of Information Act request.

Musk married, divorced, remarried, and then divorced the actress Talulah Riley. *Photograph courtesy of Talulah Riley*

Musk and Riley relax at home in Los Angeles. Musk shares the home with his five young boys. *Photograph courtesy of Talulah Riley*

photo of the three men appeared later in the press that depicted Musk with a vacant stare. "This was E's I'm-thinking-about-a-rocket-problem stance, which makes me pretty sure that he had just gotten some kind of bothersome work-related e-mail, and was clearly oblivious to the fact that a picture was being taken at all," she wrote. "This is also the reason I get suck [*sic*] a kick out of it—the spouse the camera caught is the exact spouse I encountered, say, last night en route to the bathroom, standing in the hallway frowning with his arms folded." Justine letting the world into the couple's bathroom should have served as a warning of things to come. Her blog would soon turn into one of Musk's worst nightmares.

The press had not run into a guy like Musk for a very long time. His shine as an Internet millionaire kept getting, well, shinier thanks to PayPal's ongoing success. He also had an element of mystery. There was the weird name. And there was the willingness to spend vast sums of money on spaceships and electric cars, which came across as a combination of daring, flamboyant, and downright flabbergasting. "Elon Musk has been called 'part playboy, part space cowboy,' an image hardly dispelled by a car collection that has boasted a Porsche 911 Turbo, 1967 Series 1 Jaguar, a Hamann BMW M5 plus the aforementioned McLaren F1—which he has driven at up to 215mph on a private airstrip," a British reporter gushed in 2007. "Then there was the L39 Soviet military jet, which he sold after becoming a father." The press had picked up on the fact that Musk tended to talk a huge game and then struggle to deliver on his promises in time, but they didn't much care. The game he talked was so much bigger than anyone else's that reporters were comfortable giving Musk leeway. Tesla became the darling of Silicon Valley's bloggers, who tracked its every move and were breathless in their coverage. Similarly, reporters covering SpaceX were overjoyed that a young, feisty company had arrived to needle Boeing, Lockheed, and, to a large extent, NASA. All Musk had to do was eventually bring some of these wondrous things he'd been funding to market.

While Musk put on a good show for the public and press, he'd started to get very worried about his businesses. SpaceX's second launch attempt had failed, and the reports coming in from Tesla kept getting worse. Musk had started these two adventures with a fortune nearing $200 million and had chewed through more than half the money with little to show for it. As each Tesla delay turned into a PR fiasco, the Musk glow dimmed. People in Silicon Valley began to gossip about Musk's money problems. Reporters who months earlier had been heaping adulation on Musk turned on him. The *New York Times* picked up on Tesla's transmission problems. Automotive websites griped that the Roadster might never ship. By the end of 2007, things got downright nasty. Valleywag, Silicon Valley's gossip blog, began to take a particular interest in Musk. Owen Thomas, the site's lead writer, dug into the histories of Zip2 and PayPal and played up the times Musk was ousted as CEO to undermine some of his entrepreneurial street cred. Thomas then championed the premise that Musk was a master manipulator who played fast and loose with other people's money. "It's wonderful that Musk has realized even a small part of his childhood fantasies," Thomas wrote. "But he risks destroying his dreams by refusing to reconcile them with reality." Valleywag anointed the Tesla Roadster as its No. 1 fail of 2007 among technology companies.

As his businesses and public persona suffered, Musk's home life degraded as well. His triplets—Kai, Damian, and Saxon— had arrived near the end of 2006 and joined their brothers Griffin and Xavier. According to Musk, Justine suffered from postpartum depression following the birth of the triplets. "In the spring of 2007, our marriage was having real issues," Musk

said. "It was on the rocks." Justine's blog posts back up his sentiments. She described a much less romantic Musk and felt people
treated her as "an arm ornament who couldn't possibly have
anything interesting to say" rather than as an author and her
husband's equal. During one trip to St. Barts, the Musks ended
up sharing dinner with some wealthy, influential couples. When
Justine let out her political views, one of the men at the table
made a crack about her being so opinionated. "E chuckled back,
patted my hand the way you pat a child's," Justine wrote on her
blog. From that point on, Justine ordered Musk to introduce
her as a published novelist and not just his wife and mother of
his children. The results? "E's way of doing this throughout the
rest of the trip: 'Justine wants me to tell you that she's written
novels,' which made people look at me like oh, that's just so *cute*
and didn't really help my case."

As 2007 rolled into 2008, Musk's life became more tumultuous. Tesla basically had to start over on much of the Roadster, and SpaceX still had dozens of people living in Kwajalein
awaiting the next launch of the Falcon 1. Both endeavors were
vacuuming up Musk's money. He started selling off prized pos-
sessions like the McLaren to generate extra cash. Musk tended to
shield employees from the gravity of his fiscal situation by always
encouraging them to do their best work. At the same time, he
personally oversaw all significant purchases at both companies. Musk also trained employees to make the right trade-offs
between spending money and productivity. This struck many of
the SpaceX employees as a novel idea, since they were used to
traditional aerospace companies that had huge, multiyear government contracts and no day-to-day survival pressure. "Elon
would always be at work on Sunday, and we had some chats where
he laid out his philosophy," said Kevin Brogan, the early SpaceX
employee. "He would say that everything we did was a function

of our burn rate and that we were burning through a hundred thousand dollars per day. It was this very entrepreneurial, Silicon Valley way of thinking that none of the aerospace engineers in Los Angeles were dialed into. Sometimes he wouldn't let you buy a part for two thousand dollars because he expected you to find it cheaper or invent something cheaper. Other times, he wouldn't flinch at renting a plane for ninety thousand dollars to get something to Kwaj because it saved an entire workday, so it was worth it. He would place this urgency that he expected the revenue in ten years to be ten million dollars a day and that every day we were slower to achieve our goals was a day of missing out on that money."

Musk had become all consumed with Tesla and SpaceX out of necessity, and there can be no doubt that this exacerbated the tensions in his marriage. The Musks had a team of nannies to help with their five children, but Elon could not spend much time at home. He worked seven days a week and quite often split his time between Los Angeles and San Francisco. Justine needed a change. During moments of self-reflection, she felt sickened, perceiving herself a trophy wife. Justine longed to be Elon's partner again and to feel some of that spark from their early days before life had turned so dazzling and demanding. It's not clear how much Musk let on to Justine about his dwindling bank account. She has long maintained that Musk kept her in the dark about the family's financial arrangements. But some of Musk's closest friends did get a glimpse into the worsening financial situation. In the first half of 2008, Antonio Gracias, the founder and CEO of Valor Equity, met Musk for dinner. Gracias had been an investor in Tesla and had become one of Musk's closest friends and allies, and he could see Musk agonizing over his future. "Things were starting to be difficult with Justine, but they were still together," Gracias said. "During that dinner, Elon said, 'I will spend my last dollar on these companies. If we have to move into Justine's parents' basement, we'll do it.'"

The option of moving in with Justine's parents expired on June 16, 2008, when Musk filed for divorce. The couple did not disclose the situation right away, although Justine left hints on her blog. In late June, she posted a quotation from Moby without any additional context: "There's no such thing as a well-adjusted public figure. If they were well adjusted they wouldn't try to be a public figure." The next entry had Justine house hunting for undisclosed reasons with Sharon Stone, and a couple of entries later she talked about "a major drama" that she'd been dealing with. In September, Justine wrote her first blog post explicitly about the divorce, saying, "We had a good run. We married young, took it as far as we could and now it is over." Valleywag naturally followed with a story about the divorce and noted that Musk had been seen out with a twenty-something actress.

The media coverage and divorce freed Justine to write about her private life in a much more liberated way. In the posts that followed, she gave her account of how the marriage ended, her views on Musk's girlfriend and future second wife, and the inner workings of the divorce proceedings. For the first time, the public had access to a deeply unpleasant portrayal of Musk and received some firsthand accounts—albeit from an ex-wife—of his hard-line behavior. The writing may have been biased, but it provided a window into how Musk operated. Here's one post about the lead-up to the divorce and its rapid execution:

Divorce, for me, was like the bomb you set off when all other options have been exhausted. I had not yet given up on the diplomacy option, which was why I hadn't already filed. We were still in the early stages of marital counseling (three sessions total). Elon, however, took matters into his own hands—he tends to like to do that—when he gave me an ultimatum: "Either we fix [the marriage] today, or I will divorce you tomorrow."

That night, and again the next morning, he asked me what I wanted to do. I stated emphatically that I was not ready to unleash the dogs of divorce; I suggested that "we" hold off for at least another week. Elon nodded, touched the top of my head, and left. Later that same morning I tried to make a purchase and discovered that he had cut off my credit card, which is when I also knew that he had gone ahead and filed (as it was, E did not tell me directly; he had another person do it).

For Musk, each online missive from Justine created another public relations crisis that added to the endless stream of issues faced by his companies. The image he'd sculpted over the years appeared ready to crumble alongside his businesses. It was a disaster scenario.

Soon enough, the Musks had achieved celebrity divorce status. Mainstream outlets joined Valleywag in poring over court filings tied to the breakup, particularly as Justine fought for more money. During the PayPal days, Justine had signed a postnuptial agreement and now argued that she didn't really have the time or inclination to dig into the ramifications of the paperwork. Justine took to her blog in an entry titled "golddigger," and said she was fighting for a divorce settlement that would include their house, alimony and child support, $6 million in cash, 10 percent of Musk's Tesla stock, 5 percent of Musk's SpaceX stock, and a Tesla Roadster. Justine also appeared on CNBC's show *Divorce Wars* and wrote an article for *Marie Claire* titled " 'I Was a Starter Wife': Inside America's Messiest Divorce."

The public tended to side with Justine during all of this and couldn't quite figure out why a billionaire was fighting his wife's seemingly fair requests. A major problem for Musk, of course, was that his assets were anything but liquid with most of his net worth being tied up in Tesla and SpaceX stock. The couple eventually settled with Justine getting the house, $2 million in cash (minus her legal fees), $80,000 a month in alimony and child support for seventeen years, and a Tesla Roadster.*

Years after the settlement, Justine still struggled to speak about her relationship with Musk. During our interview, she broke down in tears several times and needed moments to compose her thoughts. Musk, she said, had hidden many things from her during their marriage and ultimately treated her much like a business adversary to be conquered during the divorce. "We were at war for a while, and when you go to war with Elon, it's pretty brutal," she said. Well after their marriage ended, Justine continued to blog about Musk. She wrote about Riley and provided commentary on his parenting. One post gave Musk a hard time for banning stuffed animals from the house when their twins turned seven. Asked about this, Justine said, "Elon

---

* Musk fought to set the record straight, as he saw it, on the *Huffington Post* and wrote a 1,500-word essay. Musk maintained that two months of negotiations with independent parties had gone into the postnuptial agreement, which kept the couple's assets separate so that Musk could get the spoils from his companies and Justine could get the spoils from her books. "In mid 1999, Justine told me that if I proposed to her, she would say yes," Musk wrote. "Since this was not long after the sale of my first company, Zip2, to Compaq, and the subsequent cofounding of PayPal, friends and family advised me to separate whether the marriage was for love or money." After the settlement, Musk asked Arianna Huffington to remove his essay about the divorce from her website. "I don't want to dwell on past negativity," Musk said. "You can always find things on the Internet. So it's not like it's gone. It's just not easily found."

is hard-core. He grew up in a tough culture and tough circumstances. He had to become very tough to not only thrive but to conquer the world. He doesn't want to raise soft overprivileged kids with no direction." Comments like these seemed to indicate that Justine still admired or at least understood Musk's strong will.*

In the weeks after he first filed for divorce in mid-June of 2008, Musk tumbled into a deep funk. Bill Lee started to worry about his friend's mental state and, as one of Musk's more free-spirited friends, wanted to do something to cheer him up. Now and again, Musk and Lee, an investor, would take trips overseas and mix business and pleasure. The time was right for just such a journey, and they set off for London at the start of July.

The decompression program began poorly. Musk and Lee visited the headquarters of Aston Martin to talk to people and get a tour of the factory. One "expert" treated Musk like an amateur car builder, talking down to him and suggesting that he knew more about electric vehicles than anyone else on the planet. "He was a complete douche," as Lee put it, and the men did their best to make a hasty exit back to central London. Along the way, Musk had a nagging stomach pain turn severe. At the time, Lee was

---

* The pair have continued to have their difficulties. For a long time, Musk ran all of the child-sharing scheduling through his assistant Mary Beth Brown rather than dealing directly with Justine. "I was really pissed-off about that," Justine said. And the time Justine cried the most during our conversation came as she weighed the pros and cons of the children growing up on a grand stage where they're whisked away to the Super Bowl or Spain in a private jet on a moment's notice or asked to play at the Tesla factory. "I know the kids really look up to him," she said. "He takes them everywhere and provides a lot of experiences for them. My role as the mother is to create this reality where I provide a sense of normalcy. They are not growing up in a normal family with a normal dad. Their life with me is a lot more low-key. We value different things. I am a lot more about empathy."

married to Sarah Gore, the daughter of former vice president Al Gore, who had been a medical student, and so he called her for advice. They decided that Musk might be suffering from appendicitis, and Lee took him to a medical clinic in the middle of a shopping mall. When the tests came back negative, Lee set to work trying to goad Musk into a night on the town. "Elon didn't want to go out, and I didn't really, either," Lee said. "But I was like, 'No, come on. We're all the way here.'"

Lee coaxed Musk into going to a club called Whisky Mist, in Mayfair. People had packed the small, high-end dance spot and Musk wanted to leave after ten minutes. The well-connected Lee texted a promoter friend of his, who pulled some strings to get Musk escorted into the VIP area. The promoter then reached out to some of his prettiest friends, including a twenty-two-year-old up-and-coming actress named Talulah Riley, and they soon arrived at the club as well. Riley and her two gorgeous friends had come from a charity gala and were in full-length, flowing gowns. "Talulah was in this huge Cinderella thing," Lee said. Musk and Riley were introduced by people at the club, and he perked at the sight of her dazzling figure.

Musk and Riley sat at a table with their friends but immediately zeroed in on each other. Riley had just hit it big with her portrayal of Mary Bennet in *Pride and Prejudice* and thought of herself as quite the hotshot. The older Musk, meanwhile, took on the role of the soft-spoken, sweet engineer. He whipped out his phone and displayed photos of the Falcon 1 and Roadster, although Riley thought he had just done some work on these projects and didn't realize he ran the companies building the machines. "I remember thinking that this guy probably didn't get to talk to young actresses a lot and that he seemed quite nervous," Riley said. "I decided to be really nice to him and give him a nice evening. Little did I know that he'd spoken to a lot of pretty girls in his life."* The more Musk and Riley talked, the more Lee egged them on. It was the first time in weeks that his friend appeared happy. "His stomach didn't hurt; he's not bummed, this is great," Lee said. Despite being dressed for a fairy tale, Riley didn't fall in love with Musk at first sight. But she did become more impressed and intrigued as the night went on, particularly after the club promoter introduced Musk to a stunning model, and he politely said "Hello" and then sat right back down with Riley. "I figured he couldn't be all bad after that," said Riley, who then allowed Musk to place his hand on her knee. Musk asked Riley out to dinner the next night, and she accepted.

With her curvy figure, sultry eyes, and playful good-girl demeanor, Riley was a budding film star but didn't really act the part. She grew up in the idyllic English countryside, went to a top school, and, until a week before she met Musk, had been living at home with her parents. After the night at Whisky Mist, Riley called her family to tell them about the interesting guy she had met who builds rockets and cars. Her father used to head up the National Crime Squad and went straight to his computer to conduct a background check that illuminated Musk's resume as a married international playboy with five kids. Riley's father chided his daughter for being a fool, but she held out hope that Musk had an explanation and went to dinner with him anyway.

Musk brought Lee to the dinner, and Riley brought her friend Tamsin Egerton, also a beautiful actress. Things were cooler throughout the meal as the group dined in a depressingly empty

---

* Musk recalled their meeting as follows: "She did look great, but what was going through my mind was 'Oh, I guess they are a couple of models.' You know, you can't actually talk to most models. You just can't have a conversation. But, you know, Talulah was really interested in talking about rockets and electric cars. That was the interesting thing."

restaurant. Riley waited to see what Musk would bring up on his own. Eventually, he did announce his five sons and his pending divorce. The confession proved enough to keep Riley interested and curious about where things would lead. Following the meal, Musk and Riley broke off on their own. They went for a walk through Soho and then stopped at Cafe Boheme, where Riley, a lifelong teetotaler, sipped an apple juice. Musk kept Riley's attention, and the romance began in earnest.

The couple had lunch the next day and then went to the White Cube, a modern art gallery, and then back to Musk's hotel room. Musk told Riley, a virgin, that he wanted to show her his rockets. "I was skeptical, but he did actually show me rocket videos," she said. Once Musk went back to the United States,* they kept in touch via e-mail for a couple of weeks, and then Riley booked a flight to Los Angeles. "I wasn't even thinking girlfriend or anything like that," Riley said. "I was just having fun."

Musk had other ideas. Riley had been in California for just five days when he made his move as they lay in bed talking in a tiny room at the Peninsula hotel in Beverley Hills. "He said, 'I don't want you to leave. I want you to marry me.' I think I laughed. Then, he said, 'No. I'm serious. I'm sorry I don't have a ring.' I said, 'We can shake on it if you like.' And we did. I don't remember what I was thinking at the time, and all I can say is that I was twenty-two."

Riley had been a model daughter up to that point, never giving her parents much of anything to worry about. She did well at school, had scored some tremendous acting gigs, and had a soft, sweet personality that her friends described as Snow White brought to life. But there she was on the hotel's balcony, informing her parents that she had agreed to marry a man fourteen years

---

* He asked Riley to go with him, but she turned Musk down.

her senior, who had just filed for divorce from his first wife, had five kids and two companies, and she didn't even see how she could possibly love him after knowing him for a matter of weeks. "I think my mother had a nervous breakdown," Riley said. "But I had always been highly romantic, and it actually didn't strike me as that strange." Riley flew back to England to gather her things, and her parents flew back with her to the United States to meet Musk, who belatedly asked Riley's father for his blessing. Musk did not have his own house, which left the couple moving into a home that belonged to Musk's friend the billionaire Jeff Skoll. "I had been living there a week when this random guy walked in," Riley said. "I said, 'Who are you?' He said, 'I am the homeowner. Who are you?' I told him, and then he just walked out." Musk later proposed to Riley again on the balcony of Skoll's house, unveiling a massive ring. (He has since bought her three engagement rings, including the giant first one, an everyday ring, and one designed by Musk that has a diamond surrounded by ten sapphires.) "I remember him saying, 'Being with me was choosing the hard path.' I didn't quite understand at the time, but I do now. It's quite hard, quite the crazy ride."

Riley experienced a baptism by fire. The whirlwind romance had given her the impression that she was engaged to a world conquering, jet-setting billionaire. That was true in theory but a murkier proposition in practice. As late July rolled around, Musk could see that he had just enough cash on hand to scrape through to the end of the year. Both SpaceX and Tesla would need cash infusions at some point just to pay the employees, and it was unclear where that money would come from with the world's financial markets in disarray and investments being put on hold. If things had been going more smoothly at the companies, Musk could have felt more confident about raising money, but they were not. "He would come home every day, and there would be some calamity," Riley said. "He was under immense pressure from all quarters. It was horrendous."

SpaceX's third flight from Kwajalein jumped out as Musk's most pressing concern. His team of engineers had remained camped out on the island, preparing the Falcon 1 for another run. A typical company would focus just on the task at hand. Not SpaceX. It had shipped the Falcon 1 to Kwaj in April with one set of engineers and then put another group of engineers on a new project to develop the Falcon 9, a nine-engine rocket that would take the place of the Falcon 5 and serve as a possible replacement to the retiring space shuttle. SpaceX had yet to prove it could get to space successfully, but Musk kept positioning it to bid on big-ticket NASA contracts.*

---

* By this time, Musk had built up a reputation as the hardest-charging man in the space business. Before settling on the Falcon 9, Musk planned to build something called the BFR, a.k.a. the Big Falcon Rocket or Big Fucking Rocket. Musk wanted it to have the biggest rocket engine in history. Musk's bigger, faster mentality amused, horrified and impressed some of the suppliers that SpaceX occasionally turned to for help, like Barber-Nichols Inc., a Colorado-based maker of rocket engine turbo pumps and other aerospace machinery. A few executives at Barber-Nichols—Robert Linden, Gary Frey, and Mike Forsha—were kind enough to recount their first meeting with Musk in the middle of 2002 and their subsequent dealings with him. Here's a snippet:

"Elon showed up with Tom Mueller and started telling us it was his destiny to launch things into space at lower costs and to help us become space faring people. We thought the world of Tom but weren't quite sure whether to take Elon too seriously. They began asking us for the impossible. They wanted a turbo pump to be built in less than a year for under one million dollars. Boeing might do a project like that over five years for one hundred million. Tom told us to give it our best shot, and we built it in thirteen months. Build quick and learn quickly was Elon's philosophy. He was relentless in wanting the costs to come down. Regardless of what we showed him on paper with regard to the cost of materials, he wanted the cost lower because that was part of his business model. It could be very frustrating to work with Elon. He

On July 30, 2008, the Falcon 9 had a successful test fire in Texas with all nine of its engines lighting up and producing 850,000 pounds of thrust. Three days later, in Kwaj, SpaceX's engineers fueled up the Falcon 1 and crossed their fingers. The rocket had an air force satellite as its payload, along with a couple of experiments from NASA. All told, the cargo weighed 375 pounds.

SpaceX had been making significant changes to its rocket since the last, failed launch. A traditional aerospace company would not have wanted the added risk, but Musk insisted that SpaceX push its technology forward while at the same time trying to make it work right. Among the biggest changes for the Falcon 1 was a new version of the Merlin 1 engine that relied on a tweaked cooling system.

The first launch attempt on August 2, 2008, aborted at T minus zero seconds. SpaceX regrouped and tried to launch again the same day. This time everything seemed to be going well. The Falcon 1 soared into the sky and flew spectacularly without any indication of a problem. SpaceX employees watching a webcast of the proceedings back in California let out hoots and whistles. Then, right at the moment when the first stage and second stage were to separate, there was a malfunction. An analysis after the fact would show that the new engines had delivered an unexpected thrust during the separation process that caused the first stage to bump up into the second stage, damaging the top part of the rocket and its engine.*

---

has a singular view and doesn't deviate from that. We don't know too many people that have worked for him that are happy. That said, he has driven the cost of space down and been true to his original business plan. Boeing, Lockheed, and the rest of them have become overly cautious and spend a lot of money. SpaceX has balls."

* To provide a glimpse of how well Musk knows the rockets, here he is explaining what happened from memory six years after the fact: "It was because we had upgraded the Merlin engine to a regeneratively

The failed launch left many SpaceX employees shattered. "It was so profound seeing the energy shift over the room in the course of thirty seconds," said Dolly Singh, a recruiter at SpaceX. "It was like the worst fucking day ever. You don't usually see grown-ups weeping, but there they were. We were tired and broken emotionally." Musk addressed the workers right away and encouraged them to get back to work. "He said, 'Look. We are going to do this. It's going to be okay. Don't freak out,'" Singh recalled. "It was like magic. Everyone chilled out immediately and started to focus on figuring out what just happened and how to fix it. It went from despair to hope and focus." Musk put up a positive front to the public as well. In a statement, he said that SpaceX had another rocket waiting to attempt a fourth launch and a fifth launch planned shortly after that. "I have also given the go-ahead to begin fabrication of flight six," he said. "Falcon 9 development will also continue unabated."

In reality, the third launch was a disaster with cascading consequences. Since the second stage of the rocket did not fire properly, SpaceX never got a chance to see if it had really fixed the fuel-sloshing issues that had plagued the second flight. Many of the SpaceX engineers were confident that they had solved this problem and were anxious to get to the fourth launch, believing

---

cooled engine and the thrust transient of that engine was a few seconds longer. It was only like one percent thrust for about another 1.5 seconds. And the chamber pressure was only ten PSI, which is one percent of the total. But that's below sea level pressure. On the test stand, we didn't notice anything. We thought it was fine. We thought it was just the same as before, but actually it just had this slight difference. The ambient sea level pressure was higher at roughly fifteen PSI, which disguised some effects during the test. The extra thrust caused the first stage to continue moving after stage separation and recontact the other stage. And the upper stage then started the engine inside the interstage, which caused the plasma blowback which destroyed that upper stage."

that they had an easy answer for the recent thrust problem. For Musk, the situation seemed graver. "I was super depressed," Musk said. "If we hadn't solved the slush coupling problem on flight two, or there was just some random other thing that occurred— say a mistake in the launch process or the manufacturing process unrelated to anything previous—then game over." SpaceX simply did not have enough money to try a fifth flight. He'd put $100 million into the company and had nothing to spare because of the issues at Tesla. "Flight four was it," Musk said. If, however, SpaceX could nail the fourth flight, it would instill confidence on the part of the U.S. government and possible commercial customers, paving the way for the Falcon 9 and even more ambitious projects.

Leading up to the third launch, Musk had been his usual ultra-involved self. Anyone at SpaceX who held the launch back went onto Musk's critical-path shit list. Musk would hound the person responsible about the delays but, typically, he would also do everything in his power to help solve problems. "I was personally holding up the launch once and had to give Elon twice-daily updates about what was going on," said Kevin Brogan. "But Elon would say, 'There are five hundred people at this company. What do you need?'" One of the calls must have taken place while Musk courted Riley because Brogan remembered Musk phoning from the bathroom of a London club to find out how welding had gone on a large part of the rocket. Musk fielded another call while with Riley and had to whisper as he berated the engineers. "He's giving us the pillow talk voice, so we all have to huddle around the speakerphone, while he tells us, 'You guys need to get your shit together,'" Brogan said.

With the fourth launch, the demands and anticipation had ratcheted to the point that people started making silly mistakes. Typically, the body of the Falcon 1 rocket traveled to Kwaj via barge. This time Musk and the engineers were too excited and desperate to wait for the ocean journey. Musk rented a military cargo plane to fly the rocket body from Los Angeles to Hawaii and then on to Kwaj. This would have been a fine idea except the SpaceX engineers forgot to factor in what the pressurized plane would do to the body of the rocket, which is less than an eighth of an inch thick. As the plane started its descent into Hawaii, everyone inside of it could hear strange noises coming from the cargo hold. "I looked back and could see the stage crumpling," said Bulent Altan, the former head of avionics at SpaceX. "I told the pilot to go up, and he did." The rocket had behaved much like an empty water bottle will on a plane, with the air pressure pushing against the sides of the bottle and making it buckle. Altan calculated that the SpaceX team on the plane had about thirty minutes to do something about the problem before they would need to land. They pulled out their pocketknives and cut away the shrink wrap that held the rocket's body tight. Then they found a maintenance kit on the plane and used wrenches to open up some nuts on the rocket that would allow its internal pressure to match that of the plane's. When the plane landed, the engineers divvied up the duties of calling SpaceX's top executives to tell them about the catastrophe. It was 3 A.M. Los Angeles time, and one of the executives volunteered to deliver the horrific news to Musk. The thinking at the time was that it would take three months to repair the damage. The body of the rocket had caved in in several places, baffles placed inside the fuel tank to stop the sloshing problem had broken, and an assortment of other issues had appeared. Musk ordered the team to continue on to Kwaj and sent in a reinforcement team with repair parts. Two weeks later, the rocket had been fixed inside of the makeshift hangar. "It was like being stuck in a foxhole together," Altan said. "You weren't going to quit and leave the

SpaceX headquarters let out raucous cheers. Each milestone that followed—clearing the island, engine checks coming back good—was again met with whistles and shouts. As the first stage fell away, the second stage fired up about ninety seconds into the flight and the employees turned downright rapturous, filling the webcast with their ecstatic hollering. "Perfect," said one of the talking heads. The Kestrel engine glowed red and started its six-minute burn. "When the second stage cleared, I could finally start breathing again and my knees stopped buckling," said McLaury.

The fairing opened up around the three-minute mark and fell back toward Earth. And, finally, around nine minutes into its journey, the Falcon 1 shut down just as planned and reached orbit, making it the first privately built machine to accomplish such a feat. It took six years—about four and half more than Musk had once planned—and five hundred people to make this miracle of modern science and business happen.

Earlier in the day, Musk had tried to distract himself from the mounting pressure by going to Disneyland with his brother Kimbal and their children. Musk then had to race back to make the 4 P.M. launch and walked into SpaceX's trailer control room about two minutes before blastoff. "When the launch was successful, everyone burst into tears," Kimbal said. "It was one of the most emotional experiences I've had." Musk left the control room and walked out to the factory floor, where he received a rock star's welcome. "Well, that was freaking awesome," he said. "There are a lot of people who thought we couldn't do it—a lot actually—but as the saying goes, 'the fourth time is the charm,' right? There are only a handful of countries on Earth that have done this. It's normally a country thing, not a company thing. . . . My mind is kind of frazzled, so it's hard for me to say anything, but, man, this is definitely one of the greatest days in my life, and I think probably for most people here. We showed people we can do it. This is just the first step of many. . . . I am going to have a really great party tonight. I don't know about you guys." Mary Beth Brown then tapped Musk on the shoulder and pulled him away to a meeting.

The afterglow of this mammoth victory faded soon after the party ended, and the severity of SpaceX's financial hell became top of mind again for Musk. SpaceX had the Falcon 9 efforts to support and had also immediately green-lighted the construction of another machine—the Dragon capsule—that would be used to take supplies, and one day humans, to the International Space Station. Historically, either project would cost more than $1 billion to complete, but SpaceX would have to find a way to build both machines simultaneously for a fraction of the cost. The company had dramatically increased the rate at which it hired employees and moved into a much larger headquarters in Hawthorne, California. SpaceX had a commercial flight booked to carry a satellite into orbit for the Malaysian government, but that launch and the payment for it would not arrive until the middle of 2009. In the meantime, SpaceX simply struggled to make its payroll.

The press did not know the extent of Musk's financial woes, but they knew enough to turn detailing Tesla's precarious financial situation into a favored pastime. A website called the Truth About Cars began a "Tesla Death Watch" in May 2008 and followed up with dozens of entries throughout the year. The blog took special pleasure in rejecting the idea that Musk was a true founder of the company, presenting him as the moneyman and chairman who had more or less stolen Tesla from the genius engineer Eberhard. When Eberhard started a blog detailing the pros and cons of being a Tesla customer, the auto site was all too happy to echo his gripes. *Top Gear*, a popular British television show, ripped the Roadster's range apart, by thrashing it around their race track so hard, that they declared that it would have run out of juice after 55 miles—a claim that Musk bitterly contested. "People joke about the Tesla Death Watch and all that, but it was harsh," said Kimbal Musk. "One day there were fifty articles about how Tesla will die."

Then, in October 2008 (just a couple weeks after SpaceX's successful launch), Valleywag appeared on the scene again. First it ridiculed Musk for officially taking over as CEO of Tesla and replacing Drori, on the grounds that Musk had just lucked into his past successes. It followed that by printing a tell-all e-mail from a Tesla employee. The report said that Tesla had just gone through a round of layoffs, shut down its Detroit office, and had only $9 million left in the bank. "We have over 1,200 reservations, which manes [sic] we've taken multiples of tens of millions of cash from our customers and have spent them all," the Tesla employee wrote. "Meanwhile, we only delivered less than 50 cars. I actually talked a close friend of mine into putting down $60,000 for a Tesla Roadster. I cannot conscientiously be a bystander anymore and allow my company to deceive the public and defraud our dear customers. Our customers and the general public are the reason Tesla is so loved. The fact that they are being lied to is just wrong."*

Yes, Tesla deserved much of the negative attention. Musk, though, felt like the 2008 climate with the hatred of bankers and the rich had turned him into a particularly juicy target. "I was just getting pistol-whipped," Musk said. "There was a lot of schaden-

---

* Musk would later discover the identity of this employee in an ingenious way. He copied the text of the letter into a Word document, checked the size of the file, sent it to a printer, and looked over the logs of printer activity to find one of the same size. He could then trace that back to the person who had printed the original file. The employee wrote a letter of apology and resigned.

freude at the time, and it was bad on so many levels. Justine was torturing me in the press. There were always all these negative articles about Tesla, and the stories about SpaceX's third failure. It hurt really bad. You have these huge doubts that your life is not working, your car is not working, you're going through a divorce and all of those things. I felt like a pile of shit. I didn't think we would overcome it. I thought things were probably fucking doomed."

When Musk ran through the calculations concerning SpaceX and Tesla, it occurred to him that only one company would likely even have a chance at survival. "I could either pick SpaceX or Tesla or split the money I had left between them," Musk said. "That was a tough decision. If I split the money, maybe both of them would die. If I gave the money to just one company, the probability of it surviving was greater, but then it would mean certain death for the other company. I debated that over and over." While Musk meditated on this, the economy worsened quickly and so too did Musk's financial condition. As 2008 came to an end, Musk had run out of money.

Riley began to see Musk's life as a Shakespearean tragedy. Sometimes Musk would open up to her about the issues, and other times he retreated into himself. Riley spied on Musk while he read e-mail and watched him grimace as the bad news poured in. "You'd witness him having these conversations in his head," she said. "It's really hard to watch someone you love struggle like that." Because of the long hours that he worked and his eating habits, Musk's weight fluctuated wildly. Bags formed under his eyes, and his countenance started to resemble that of a shattered runner at the back end of an ultra-marathon. "He looked like death itself," Riley said. "I remember thinking this guy would have a heart attack and die. He seemed like a man on the brink." In the middle of the night, Musk would have nightmares and yell out. "He was in physical pain," Riley said. "He would climb on me and start screaming while still asleep." The couple had to start borrowing hundreds of thousands of dollars from Musk's friend Skoll, and Riley's parents offered to remortgage their house. Musk no longer flew his jet back and forth between Los Angles and Silicon Valley. He took Southwest.

Burning through about $4 million a month, Tesla needed to close another major round of funding to get through 2008 and stay alive. Musk had to lean on friends just to try to make payroll from week to week, as he negotiated with investors. He sent impassioned pleas to anyone he could think of who might be able to spare some money. Bill Lee invested $2 million in Tesla, and Sergey Brin invested $500,000. "A bunch of Tesla employees wrote checks to keep the company going," said Diarmuid O'Connell, the vice president of business development at Tesla. "They turned into investments, but, at the time, it was twenty-five or fifty thousand dollars that you didn't expect to see again. It just seemed like holy shit, this thing is going to crater." Kimbal had lost most of his money during the recession when his investments bottomed out but sold what he had left and put it into Tesla as well. "I was close to bankruptcy," Kimbal said. Tesla had set the prepayments that customers made for the Roadsters aside, but Musk now needed to use that money to keep the company going and soon those funds were gone, too. These fiscal maneuvers worried Kimbal. "I'm sure Elon would have found a way to make things right, but he definitely took risks that seemed like they could have landed him in jail for using someone else's money," he said.

In December 2008, Musk mounted simultaneous campaigns to try to save his companies. He heard a rumor that NASA was on the verge of awarding a contract to resupply the space station. SpaceX's fourth launch had put it in a position to receive some of this money, which was said to be in excess of $1 billion.

Musk reached out through some back channels in Washington and found out that SpaceX might even be a front-runner for the deal. Musk began doing everything in his power to assure people that the company could meet the challenge of getting a capsule to the ISS. As for Tesla, Musk had to go to his existing investors and ask them to pony up for another round of funding that needed to close by Christmas Eve to avoid bankruptcy. To give the investors some measure of confidence, Musk made a last-ditch effort to raise all the personal funds he could and put them into the company. He took out a loan from SpaceX, which NASA approved, and earmarked the money for Tesla. Musk went to the secondary markets to try to sell some of his shares in SolarCity. He also seized about $15 million that came through when Dell acquired a data center software start-up called Everdream, founded by Musk's cousins, in which he had invested. "It was like the fucking Matrix," Musk said, describing his financial maneuvers. "The Everdream deal really saved my butt."

Musk had cobbled together $20 million, and asked Tesla's existing investors—since no new investors materialized—to match that figure. The investors agreed, and on December 3, 2008, they were in the process of finalizing the paperwork for the funding round when Musk noticed a problem. VantagePoint Capital Partners had signed all of the paperwork except for one crucial page. Musk phoned up Alan Salzman, VantagePoint's cofounder and managing partner, to ask about the situation. Salzman informed Musk that the firm had a problem with the investment round because it undervalued Tesla. "I said, 'I've got an excellent solution then. Take my entire portion of the deal. I had a real hard time coming up with the money. Based on the cash we have in the bank right now, we will bounce payroll next week. So unless you've got another idea, can you either just participate as much as you'd like, or allow the round to go through because otherwise we will be bankrupt.'" Salzman balked and told Musk to come in the following week at 7 A.M. to present to VantagePoint's top brass. Not having a week of time to work with, Musk asked to come in the next day, and Salzman refused that offer, forcing Musk to continue taking on loans. "The only reason he wanted the meeting at his office was for me to come on bended knee begging for money so he could say, 'No,'" Musk theorized.

VantagePoint declined to speak about this period, but Musk believed that Salzman's tactics were part of a mission to bankrupt Tesla. Musk feared that VantagePoint would oust him as CEO, recapitalize Tesla, and emerge as the major owner of the carmaker. It could then sell Tesla to a Detroit automaker or focus on selling electric drivetrains and battery packs instead of making cars. Such reasoning would have been quite practical from a business standpoint but did not match up with Musk's goals for Tesla. "VantagePoint was forcing that wisdom down the throat of an entrepreneur who wanted to do something bigger and bolder," said Steve Jurvetson, a partner at Draper Fisher Jurvetson and Tesla investor. "Maybe they're used to a CEO buckling, but Elon doesn't do that." Instead, Musk took another huge risk. Tesla recharacterized the funding as a debt round rather than an equity round, knowing that VantagePoint could not interfere with a debt deal. The tricky part of this strategy was that investors like Jurvetson who wanted to help Tesla were put in a bind because venture capital firms are not structured to do debt deals, and convincing their backers to alter their normal rules of engagement for a company that could very well go bankrupt in a matter of days would be a very tough ask. Knowing this, Musk bluffed. He told the investors that he would take another loan from SpaceX and fund the entire round—all $40 million—himself. The tactic worked. "When you have scarcity, it naturally reinforces greed and leads to more interest," Jurvetson said. "It was also easier for us to go back to our firms and say, 'Here is the deal. Go or no go?'" The deal ended up closing on Christmas Eve, hours before Tesla would have gone bankrupt. Musk had just a few hundred thousand dollars left and could not have made payroll the next day. Musk ultimately put in $12 million, and the investment firms put up the rest.

At SpaceX, Musk and the company's top executives had spent most of December in a state of fear. According to reports in the press, SpaceX, the onetime front-runner for the large NASA contract, had suddenly lost favor with the space agency. Michael Griffin, who had once almost been a cofounder of SpaceX, was the head of NASA and had turned on Musk. Griffin did not care for Musk's aggressive business tactics, seeing him as borderline unethical. Others have suggested that Griffin ended up being jealous of Musk and SpaceX.* On December 23, 2008, however, SpaceX received a shock. People inside NASA had backed SpaceX to become a supplier for the ISS. The company received $1.6 billion as payment for twelve flights to the space station. Staying with Kimbal in Boulder, Colorado, for the holidays, Musk broke down in tears as the SpaceX and Tesla transactions processed. "I hadn't had an opportunity to buy a Christmas present for Talulah or anything," he said. "I went running down the fucking street in Boulder, and the only place that was open sold these shitty trinkets, and they were about to close. The best thing I could find were these plastic monkeys with coconuts—those 'see no evil, hear no evil' monkeys."

----

* Griffin had pined to build a massive new spacecraft that would also solidify his mark on the industry. But, with the election of Barack Obama in 2008, the Bush appointee knew that his time as NASA chief was coming to an end and that SpaceX appeared poised to build the most interesting machines moving forward.

For Gracias, the Tesla and SpaceX investor and Musk's friend, the 2008 period told him everything he would ever need to know about Musk's character. He saw a man who arrived in the United States with nothing, who had lost a child, who was being pilloried in the press by reporters and his ex-wife and who verged on having his life's work destroyed. "He has the ability to work harder and endure more stress than anyone I've ever met," Gracias said. "What he went through in 2008 would have broken anyone else. He didn't just survive. He kept working and stayed focused." That ability to stay focused in the midst of a crisis stands as one of Musk's main advantages over other executives and competitors. "Most people who are under that sort of pressure fray," Gracias said. "Their decisions go bad. Elon gets hyperrational. He's still able to make very clear, long-term decisions. The harder it gets, the better he gets. Anyone who saw what he went through firsthand came away with more respect for the guy. I've just never seen anything like his ability to take pain."

9

LIFTOFF

THE FALCON 9 HAS BECOME SPACEX'S WORKHORSE. THE ROCKET LOOKS— let's face it—like a giant white phallus. It stands 224.4 feet tall, is 12 feet across, and weighs 1.1 million pounds. The rocket is powered by nine engines arranged in an "octaweb" pattern at its base with one engine in the center and eight others encircling it. The engines connect to the first stage, or the main body of the rocket, which bears the blue SpaceX insignia and an American flag. The shorter second stage of the rocket sits on top of the first and is the one that actually ends up doing things in space. It can be outfitted with a rounded container for carrying satellites or a capsule capable of transporting humans. By design, there's nothing particularly flashy about the Falcon 9's outward appearance. It's the spaceship equivalent of an Apple laptop or a Braun kettle—an elegant, purposeful machine stripped of frivolity and waste.

SpaceX sometimes uses Vandenberg Air Force Base in Southern California to send up these Falcon 9 rockets. Were it not owned by the military, the base would be a resort. The Pacific Ocean runs for miles along its border, and its grounds have wide-open shrubby fields dotted by green hills. Nestled into one hilly spot just at the ocean's edge are a handful of launchpads. On launch days, the white Falcon 9 breaks up the blue and green landscape, pointing skyward and leaving no doubt about its intentions.

About four hours before a launch, the Falcon 9 starts getting filled with an immense amount of liquid oxygen and rocket-grade kerosene. Some of the liquid oxygen vents out of the rocket as it awaits launch and is kept so cold that it boils off on contact with the metal and air, forming white plumes that stream down the rocket's sides. This gives the impression of the Falcon 9 huffing and puffing as it limbers up before the journey. The engineers inside of SpaceX's mission control monitor these fuel systems and all manner of other items. They chat back and forth through headsets and begin cycling through their launch checklist, consumed by what people in the business call "go fever" as they move from one approval to the next. Ten minutes before launch, the humans step out of the way and leave the remaining processes up to automated machines. Everything goes quiet, and the tension builds until right before the main event. That's when, out of nowhere, the Falcon 9 breaks the silence by letting out a loud gasp.

A white latticed support structure pulls away from its body. The T-minus-ten-seconds countdown begins. Nothing much happens from ten down to four. At the count of three, however, the engines ignite, and the computers conduct a last, oh-so-rapid, health check. Four enormous metal clamps hold the rocket down, as computing systems evaluate all nine engines and measure if there's sufficient downward force being produced. By the time zero arrives, the rocket has decided that all is well enough to go through with its mission, and the clamps release. The rocket goes to war with inertia, and then, with flames surrounding its base and snow-thick plumes of the liquid oxygen filling the air, it shoots up. Seeing something so large hold so straight and steady while suspended in midair is hard for the brain to register. It is foreign, inexplicable. About twenty seconds after liftoff, the spectators placed safely a few miles away catch the first faceful of the Falcon 9's rumble. It's a distinct sound—a sort of staccato crackling that arises from chemicals whipped into a violent frenzy. Pant legs vibrate from shock waves produced by a stream of sonic booms coming out of the Falcon 9's exhaust. The white rocket climbs higher and higher with impressive stamina. After about a minute, it's just a red spot in the sky, and then—poof—it's gone. Only a cynical dullard could come away from witnessing this feeling anything other than wonder at what man can accomplish.

For Elon Musk, this spectacle has turned into a familiar experience. SpaceX has metamorphosed from the joke of the aeronautics industry into one of its most consistent operators. SpaceX sends a rocket up about once a month, carrying satellites for companies and nations and supplies to the International Space Station. Where the Falcon 1 blasting off from Kwajalein was the work of a start-up, the Falcon 9 taking off from Vandenberg is the work of an aerospace superpower. SpaceX can undercut its U.S. competitors—Boeing, Lockheed Martin, Orbital Sciences—on price by a ridiculous margin. It also offers U.S. customers a peace of mind that its rivals can't. Where these competitors rely on Russian and other foreign suppliers, SpaceX makes all of its machines from scratch in the United States. Because of its low costs, SpaceX has once again made the United States a player in the worldwide commercial launch market. Its $60 million per launch cost is much less than what Europe and Japan charge and trumps even the relative bargains offered by the Russians and Chinese, who have the added benefit of decades of sunk government investment into their space programs as well as cheap labor.

The United States continues to take great pride in having Boeing compete against Airbus and other foreign aircraft makers. For some reason, though, government leaders and the public have been willing to concede much of the commercial launch market. It's a disheartening and shortsighted position. The total market for satellites, related services, and the rocket launches needed to carry them to space has exploded over the past decade from about $60 billion per year to more than $200 billion.[11] A number of countries pay to send up their own spy, communication, and weather satellites. Companies then turn to space for television, Internet, radio, weather, navigation, and imaging services. The machines in space supply the fabric of modern life, and they're going to become more capable and interesting at a rapid pace. A whole new breed of satellite makers has just appeared on the scene with the ability to answer Google-like queries about our planet. These satellites can zoom in on Iowa and determine when cornfields are at peak yields and ready to harvest, and they can count cars in Wal-Mart parking lots throughout California to calculate shopping demand during the holiday season. The start-ups making these types of innovative machines must often turn to the Russians to get them into space, but SpaceX intends to change that.

The United States has remained competitive in the most lucrative parts of the space industry, building the actual satellites and complementary systems and services to run them. Each year, the United States makes about one-third of all satellites and takes about 60 percent of the global satellite revenue. The majority of this revenue comes from business done with the U.S. government. China, Europe, and Russia account for almost all of the remaining satellite sales and launches. It's expected that China's role in the space industry will increase, while Russia has vowed to spend $50 billion on revitalizing its space program. This leaves the United States dealing with two of its least-favored nations in space matters and doing so without much leverage. Case in point: the retirement of the space shuttle made the United States totally dependent on the Russians to get astronauts to the ISS. Russia gets to charge $70 million per person for the trip and to cut the United States off as it sees fit during political rifts. At present, SpaceX looks like the best hope of breaking this cycle and giving back to America its ability to take people into space.

SpaceX has become the free radical trying to upend everything about this industry. It doesn't want to handle a few launches per year or to rely on government contracts for survival. Musk's goal is to use manufacturing breakthroughs and launchpad advances to create a drastic drop in the cost of getting things to space. Most significant, he's been testing rockets that can push their payload to space and then return to Earth and land with supreme accuracy on a pad floating at sea or even their original launchpad. Instead of having its rockets break apart after crashing into the sea, SpaceX will use reverse thrusters to lower them down softly and reuse them. Within the next few years, SpaceX expects to cut its price to at least one-tenth that of its rivals. Reusing its rockets will drive the bulk of this reduction and SpaceX's competitive advantage. Imagine one airline that flies the same plane over and over again, competing against others that dispose of their planes after every flight.* Through its cost advantages,

---

* It should be noted that there are many people in the space industry who doubt reusable rockets will work, in large part because of the stress the machines and metal go through during launch. It's not clear that the most prized customers will even consider the reused spacecraft for launches due to their inherent risks. This is a big reason that other

SpaceX hopes to take over the majority of the world's commercial launches, and there's evidence that the company is on its way toward doing just that. To date, it has flown satellites for Canadian, European, and Asian customers and completed about two dozen launches. Its public launch manifest stretches out for a number of years, and SpaceX has more than fifty flights planned, which are all together worth more than $5 billion. The company remains privately owned with Musk as the largest shareholder alongside outside investors including venture capital firms like the Founders Fund and Draper Fisher Jurvetson, giving it a competitive ethos its rivals lack. Since getting past its near-death experience in 2008, SpaceX has been profitable and is estimated to be worth $12 billion.

Zip2, PayPal, Tesla, SolarCity—they are all expressions of Musk. SpaceX is Musk. Its foibles emanate directly from him, as do its successes. Part of this comes from Musk's maniacal attention to detail and involvement in every SpaceX endeavor. He's hands-on to a degree that would make Hugh Hefner feel inadequate. Part of it stems from SpaceX being the apotheosis of the Cult of Musk. Employees fear Musk. They adore Musk. The give up their lives for Musk, and they usually do all of this simultaneously.

Musk's demanding management style can only flourish because of the otherworldly—in a literal sense—aspirations of the company. While the rest of the aerospace industry has been content to keep sending what look like relics from the 1960s into space, SpaceX has made a point of doing just the opposite. Its reusable rockets and reusable spaceships look like true twenty-

---

countries and companies have not pursued the technology. There's a camp of space experts who think Musk is flat-out wasting his time, and that engineering calculations already prove the reusable rockets to be a fool's errand.

first-century machines. The modernization of the equipment is not just for show. It reflects SpaceX's constant push to advance its technology and change the economics of the industry. Musk does not simply want to lower the cost of deploying satellites and resupplying the space station. He wants to lower the cost of launches to the point that it becomes economical and practical to fly thousands upon thousands of supply trips to Mars and start a colony. Musk wants to conquer the solar system, and, as it stands, there's just one company where you can work if that sort of quest gets you out of bed in the morning.

It seems unfathomable, but the rest of the space industry has made space boring. The Russians, who dominate much of the business of sending things and people to space, do so with decades-old equipment. The cramped Soyuz capsule that takes people to the space station has mechanical knobs and computer screens that appear unchanged from its inaugural 1966 flight. Countries new to the space race have mimicked the antiquated Russian and American equipment with maddening accuracy. When young people get into the aerospace industry, they're forced to either laugh or cry at the state of the machines. Nothing sucks the fun out of working on a spaceship like controlling it with mechanisms last seen in a 1960s laundromat. And the actual work environment is as outmoded as the machines. Hotshot college graduates have historically been forced to pick between a variety of slow-moving military contractors and interesting but ineffectual start-ups.

Musk has managed to take these negatives surrounding the aerospace business and turn them into gains for SpaceX. He's presented the company as anything but another aerospace contractor. SpaceX is the hip, forward-thinking place that's brought the perks of Silicon Valley—namely frozen yogurt, stock options, speedy decision making, and a flat corporate structure—to a staid industry. People who know Musk well tend to describe him more as a general than a CEO, and this is apt. He's built an engineering army by having the pick of just about anyone in the business that SpaceX wants.

The SpaceX hiring model places some emphasis on getting top marks at top schools. But most of the attention goes toward spotting engineers who have exhibited type A personality traits over the course of their lives. The company's recruiters look for people who might excel at robot-building competitions or who are car-racing hobbyists who have built unusual vehicles. The object is to find individuals who ooze passion, can work well as part of a team, and have real-world experience bending metal. "Even if you're someone who writes code for your job, you need to understand how mechanical things work," said Dolly Singh, who spent five years as the head of talent acquisition at SpaceX. "We were looking for people that had been building things since they were little."

Sometimes these people walked through the front door. Other times, Singh relied on a handful of enterprising techniques to find them. She became famous for trawling through academic papers to find engineers with very specific skills, cold-calling researchers at labs and plucking possessed engineers out of college. At trade shows and conferences, SpaceX recruiters wooed interesting candidates they had spotted with a cloak-and-dagger shtick. They would hand out blank envelopes that contained invitations to meet at a specific time and place, usually a bar or restaurant near the event, for an initial interview. The candidates that showed up would discover they were among only a handful of people who been anointed out of all the conference attendees. They were immediately made to feel special and inspired.

Like many tech companies, SpaceX subjects potential hires to a gauntlet of interviews and tests. Some of the interviews are easygoing chats in which both parties get to feel each other out; others are filled with quizzes that can be quite hard. Engineers tend to face the most rigorous interrogations, although business types and salesmen are made to suffer, too. Coders who expect to pass through standard challenges have rude awakenings. Companies will typically challenge software developers on the spot by asking them to solve problems that require a couple of dozen lines of code. The standard SpaceX problem requires five hundred or more lines of code. All potential employees who make their way to the end of the interview process then handle one more task. They're asked to write an essay for Musk about why they want to work at SpaceX.

The reward for solving the puzzles, acting clever in interviews, and penning up a good essay is a meeting with Musk. He interviewed almost every one of SpaceX's first one thousand hires, including the janitors and technicians, and has continued to interview the engineers as the company's workforce swelled. Each employee receives a warning before going to meet with Musk. The interview, he or she is told, could last anywhere from thirty seconds to fifteen minutes. *Elon will likely keep on writing e-mails and working during the initial part of the interview and not speak much. Don't panic. That's normal. Eventually, he will turn around in his chair to face you. Even then, though, he might not make actual eye contact with you or fully acknowledge your presence. Don't panic. That's normal. In due course, he will speak to you.* From that point, the tales of engineers who have interviewed with Musk run the gamut from torturous experiences to the sublime. He might ask one question or he might ask several. You can be sure, though, that he will roll out the Riddle: "You're standing on the surface of the Earth. You walk one mile south, one mile west, and one mile north. You end up exactly where you started. Where are you?" One answer to that is the North Pole, and most

of the engineers get it right away. That's when Musk will follow with "Where else could you be?" The other answer is somewhere close to the South Pole where, if you walk one mile south,
the circumference of the Earth becomes one mile. Fewer engineers get this answer, and Musk will happily walk them through
that riddle and others and cite any relevant equations during his
explanations. He tends to care less about whether or not the person gets the answer than about how they describe the problem
and their approach to solving it.

When speaking to potential recruits, Singh tried to energize
them and be up front about the demands of SpaceX and of Musk
at the same time. "The recruiting pitch was SpaceX is special
forces," she said. "If you want as hard as it gets, then great. If
not, then you shouldn't come here." Once at SpaceX, the new
employees found out very quickly if they were indeed up for the
challenge. Many of them would quit within the first few months
because of the ninety-plus-hour workweeks. Others quit because
they could not handle just how direct Musk and the other executives were during meetings. "Elon doesn't know about you and
he hasn't thought through whether or not something is going to
hurt your feelings," Singh said. "He just knows what the fuck he
wants done. People who did not normalize to his communication
style did not do well."

There's an impression that SpaceX suffers from incredibly
high turnover, and the company has without question churned
through a fair number of bodies. Many of the key executives who
helped start the company, however, have hung on for a decade or
more. Among the rank-and-file engineers, most people stay on
for at least five years to have their stock options vest and to see
their projects through. This is typical behavior for any technology company. SpaceX and Musk also seem to inspire an unusual
level of loyalty. Musk has managed to conjure up that Steve Jobs–

like zeal among his troops. "His vision is so clear," Singh said. "He almost hypnotizes you. He gives you the crazy eye, and it's like, yes, we can get to Mars." Take that a bit further and you arrive at a pleasure-pain, sadomasochistic vibe that comes with working for Musk. Numerous people interviewed for this book decried the work hours, Musk's blunt style, and his sometimes ludicrous expectations. Yet almost every person—even those who had been fired—still worshipped Musk and talked about him in terms usually reserved for superheroes or deities.

SpaceX's original headquarters in El Segundo were not quite up to the company's desired image as a place where the cool kids want to work. This is not a problem for SpaceX's new facility in Hawthorne. The building's address is 1 Rocket Road, and it has the Hawthorne Municipal Airport and several tooling and manufacturing companies as neighbors. While the SpaceX building resembles the others in size and shape, its all-white color makes it the obvious outlier. The structure looks like a gargantuan, rectangular glacier that's been planted in the midst of a particularly soulless portion of Los Angeles County's sprawl.

Visitors to SpaceX have to walk past a security guard and through a small executive parking lot where Musk parks his black Model S, which flanks the building's entryway. The front doors are reflective and hide what's on the inside, which is more white. There are white walls in the foyer, a funky white table in the waiting area, and a white check-in desk with a pair of orchids sitting in white pots. After going through the registration process, guests are given a name badge and led into the main SpaceX office space. Musk's cubicle—a supersize unit—sits to the right where he has a couple of celebratory *Aviation Week* magazine covers up on the wall, pictures of his boys, next to a huge flat-screen monitor, and various knickknacks on his desk, including a boomerang, some books, a bottle of wine, and a giant samurai

sword named Lady Vivamus, which Musk received when he won
the Heinlein Prize, an award given for big achievements in commercial space. Hundreds of other people work in cubicles amid
the big, wide-open area, most of them executives, engineers,
software developers, and salespeople tapping away on their computers. The conference rooms that surround their desks all have
space-themed names like Apollo or Wernher von Braun and little
nameplates that explain the label's significance. The largest conference rooms have ultramodern chairs—high-backed, sleek red
jobs that surround large glass tables—while panoramic photos of
a Falcon 1 taking off from Kwaj or the Dragon capsule docking
with the ISS hang on the walls in the background.

Take away the rocket swag and the samurai sword and this
central part of the SpaceX office looks just like what you might
find at your run-of-the-mill Silicon Valley headquarters. The
same thing cannot be said for what visitors encounter as they
pass through a pair of double doors into the heart of the SpaceX
factory.

The 550,000-square-foot factory floor is difficult to process at
first glance. It's one continuous space with grayish epoxied floors,
white walls, and white support columns. A small city's worth of
stuff—people, machines, noise—has been piled into this area. Just
near the entryway, one of the Dragon capsules that has gone to the
ISS and returned to Earth hangs from the ceiling with black burn
marks running down its side. Just under the capsule on the ground
are a pair of the twenty-five-foot-long landing legs built by SpaceX
to let the Falcon rocket come to a gentle rest on the ground after
a flight so it can be flown again. To the left side of this entryway
area there's a kitchen, and to the right side there's the mission control room. It's a closed-off area with expansive glass windows and
fronted by wall-size screens for tracking a rocket's progress. It has
four rows of desks with about ten computers each for the mission

control staff. Step a bit farther into the factory and there are a handful of industrial work areas separated from each other in the most informal of ways. In some spots there are blue lines on the floor to mark off an area and in other spots blue workbenches have been arranged in squares to cordon off the space. It's a common sight to have one of the Merlin engines raised up in the middle of one of these work areas with a half dozen technicians wiring it up and tuning its bits and pieces.

Just behind these workspaces is a glass-enclosed square big enough to fit two of the Dragon capsules. This is a clean room where people must wear lab coats and hairnets to fiddle with the capsules without contaminating them. About forty feet to the left, there are several Falcon 9 rockets lying next to each other horizontally that have been painted and await transport. There are some areas tucked in between all of this that have blue walls and appear to have been covered by fabric. These are top-secret zones where SpaceX might be working on a fanciful astronaut's outfit or rocket part that it has to hide from visitors and employees not tied to the projects. There's a large area off to the side where SpaceX builds all of its electronics, another area for creating specialized composite materials, and another for making the bus-sized fairings that wrap around the satellites. Hundreds of people move about at the same time through the factory—a mix of gritty technicians with tattoos and bandanas, and young, white-collar engineers. The sweaty smell of kids who have just come off the playground permeates the building and hints at its nonstop activity.

Musk has left his personal touches throughout the factory. There are small things like the data center that has been bathed in blue lights to give it a sci-fi feel. The refrigerator-sized computers under the lights have been labeled with big block letters to make it look like they were made by Cyberdyne Systems, the fictional company from the Terminator movie franchise. Near the elevators, Musk has placed a glowing life-size Iron Man figure. Surely the factory's most Muskian element is the office space that has been built smack-dab in its center. This is a three-story glass structure with meeting rooms and desks that rises up between various welding and construction areas. It looks and feels bizarre to have a see-through office inside this hive of industry. Musk, though, wanted his engineers to watch what was going on with the machines at all times and to make sure they had to walk through the factory and talk to the technicians on the way to their desks.

The factory is a temple devoted to what SpaceX sees as its major weapon in the rocket-building game, in-house manufacturing. SpaceX manufactures between 80 percent and 90 percent of its rockets, engines, electronics, and other parts. It's a strategy that flat-out dumbfounds SpaceX's competitors, like United Launch Alliance, or ULA, which openly brags about depending on more than 1,200 suppliers to make its end products. (ULA, a partnership between Lockheed Martin and Boeing, sees itself as an engine of job creation rather than a model of inefficiency.)

A typical aerospace company comes up with the list of parts that it needs for a launch system and then hands off their design and specifications to myriad third parties who then actually build the hardware. SpaceX tends to buy as little as possible to save money and because it sees depending on suppliers—especially foreign ones—as a weakness. This approach comes off as excessive at first blush. Companies have made things like radios and power distribution units for decades. Reinventing the wheel for every computer and machine on a rocket could introduce more chances for error and, in general, be a waste of time. But for SpaceX, the strategy works. In addition to building its own engines, rocket bodies, and capsules, SpaceX designs its own motherboards and circuits, sensors to detect vibrations, flight computers, and solar panels. Just by streamlining a radio, for instance, SpaceX's engineers have found that they can reduce the weight of the device by about 20 percent. And the cost savings for a homemade radio are dramatic, dropping from between $50,000 to $100,000 for the industrial-grade equipment used by aerospace companies to $5,000 for SpaceX's unit.

It's hard to believe these kinds of price differentials at first, but there are dozens if not hundreds of places where SpaceX has secured such savings. The equipment at SpaceX tends to be built out of readily available consumer electronics as opposed to "space grade" equipment used by others in the industry. SpaceX has had to work for years to prove to NASA that standard electronics have gotten good enough to compete with the more expensive, specialized gear trusted in years past. "Traditional aerospace has been doing things the same way for a very, very long time," said Drew Eldeen, a former SpaceX engineer. "The biggest challenge was convincing NASA to give something new a try and building a paper trail that showed the parts were high enough quality." To prove that it's making the right choice to NASA and itself, SpaceX will sometimes load a rocket with both the standard equipment and prototypes of its own design for testing during flight. Engineers then compare the performance characteristics of the devices. Once a SpaceX design equals or outperforms the commercial products, it becomes the de facto hardware.

There have also been numerous times when SpaceX has done pioneering work on advancing very complex hardware systems. A classic example of this is one of the factory's weirder-looking contraptions, a two-story machine designed to perform what's known as friction stir welding. The machine allows SpaceX to automate the welding process for massive sheets of metal like the ones that make up the bodies of the Falcon rockets. An arm takes one of the rocket's body panels, lines it up against another body panel, and then joins them together with a weld that could run twenty feet or more. Aerospace companies typically try to avoid welds whenever possible because they create weaknesses in the metal, and that's limited the size of metal sheets they can use and forced other design constraints. From the early days of SpaceX, Musk pushed the company to master friction stir welding, in which a spinning head is smashed at high speeds into the join between two pieces of metal in a bid to make their crystalline structures merge. It's as if you heated two sheets of aluminum foil and then joined them by putting your thumb down on the seam and twisting the metal together. This type of welding tends to result in much stronger bonds than traditional welds. Companies had performed friction stir welding before but not on structures as large as a rocket's body or to the degree to which SpaceX has used the technique. As a result of its trials and errors, SpaceX can now join large, thin sheets of metal and shave hundreds of pounds off the weight of the Falcon rockets, as it's able to use lighter-weight alloys and avoid using rivets, fasteners, and other support structures. Musk's competitors in the auto industry might soon need to do the same because SpaceX has transferred some of the equipment and techniques to Tesla. The hope is that Tesla will be able to make lighter, stronger cars.

The technology has proven so valuable that SpaceX's competitors have started to copy it and have tried to poach some of the company's experts in the field. Blue Origin, Jeff Bezos's secretive rocket company, has been particularly aggressive, hiring away Ray Miryekta, one of the world's foremost friction stir welding experts and igniting a major rift with Musk. "Blue Origin does these surgical strikes on specialized talent* offering like double their salaries. I think it's unnecessary and a bit rude,"

---

* Blue Origin also hired away a large chunk of SpaceX's propulsion team.

Musk said. Within SpaceX, Blue Origin is mockingly referred to as BO and at one point the company created an e-mail filter to detect messages with "blue" and "origin" to block the poaching. The relationship between Musk and Bezos has soured, and they no longer chat about their shared ambition of getting to Mars. "I do think Bezos has an insatiable desire to be King Bezos," Musk said. "He has a relentless work ethic and wants to kill everything in e-commerce. But he's not the most fun guy, honestly."*

In the early days of SpaceX, Musk knew little about the machines and amount of grunt work that goes into making rockets. He rebuffed requests to buy specialized tooling equipment, until the engineers could explain in clear terms why they needed certain things and until experience taught him better. Musk also had yet to master some of the management techniques for which he would become both famous and to some degree infamous.

Musk's growth as a CEO and rocket expert occurred alongside SpaceX's maturation as a company. At the start of the Falcon 1 journey, Musk was a forceful software executive trying to learn some basic things about a very different world. At Zip2 and PayPal, he felt comfortable standing up for his positions and directing teams of coders. At SpaceX, he had to pick things up on the job. Musk initially relied on textbooks to form the bulk of his rocketry knowledge. But as SpaceX hired one brilliant person after another, Musk realized he could tap into their stores of

---

* Musk has taken exception to Blue Origin and Bezos filing for patents around reusable rocket technology as well. "His patent is completely ridiculous," Musk said. "People have proposed landing on a floating platform in the ocean for a half century. There's no chance whatsoever of the patent being upheld because there's five decades of prior art of people who proposed that six ways to Sunday in fiction and nonfiction. It's like Dr. Seuss, green eggs and fucking ham. That's how many ways it's been proposed. The issue is doing it and like actually creating a rocket that can make that happen."

knowledge. He would trap an engineer in the SpaceX factory and set to work grilling him about a type of valve or specialized material. "I thought at first that he was challenging me to see if I knew my stuff," said Kevin Brogan, one of the early engineers. "Then I realized he was trying to learn things. He would quiz you until he learned ninety percent of what you know." People who have spent significant time with Musk will attest to his abilities to absorb incredible quantities of information with near-flawless recall. It's one of his most impressive and intimidating skills and seems to work just as well in the present day as it did when he was a child vacuuming books into his brain. After a couple of years running SpaceX, Musk had turned into an aerospace expert on a level that few technology CEOs ever approach in their respective fields. "He was teaching us about the value of time, and we were teaching him about rocketry," Brogan said.

In regards to time, Musk may well set more aggressive delivery targets for very difficult-to-make products than any executive in history. Both his employees and the public have found this to be one of the more jarring aspects of Musk's character. "Elon has always been optimistic," Brogan said. "That's the nice word. He can be a downright liar about when things need to get done. He will pick the most aggressive time schedule imaginable assuming everything goes right, and then accelerate it by assuming that everyone can work harder."

Musk has been pilloried by the press for setting and then missing product delivery dates. It's one of the habits that got him in the most trouble as SpaceX and Tesla tried to bring their first products to market. Time and again, Musk found himself making a public appearance where he had to come up with a new batch of excuses for a delay. Reminded about the initial 2003 target date to fly the Falcon 1, Musk acted shocked. "Are you serious?" he said. "We said that? Okay, that's ridiculous. I think I just didn't know what the hell I was talking about. The only thing I had prior experience in was software, and, yeah, you can write a bunch of software and launch a website in a year. No problem. This isn't like software. It doesn't work that way with rockets." Musk simply cannot help himself. He's an optimist by nature, and it can feel like he makes calculations for how long it will take to do something based on the idea that things will progress without flaw at every step and that all the members of his team have Muskian abilities and work ethics. As Brogan joked, Musk might forecast how long a software project will take by timing the amount of seconds needed physically to write a line of code and then extrapolating that out to match however many lines of code he expects the final piece of software to be. It's an imperfect analogy but one that does not seem that far off from Musk's worldview. "Everything he does is fast," Brogan said. "He pees fast. It's like a fire hose—three seconds and out. He's authentically in a hurry."

Asked about his approach, Musk said,

> I certainly don't try to set impossible goals. I think impossible goals are demotivating. You don't want to tell people to go through a wall by banging their head against it. I don't ever set intentionally impossible goals. But I've certainly always been optimistic on time frames. I'm trying to recalibrate to be a little more realistic.
>
> I don't assume that it's just like 100 of me or something like that. I mean, in the case of the early SpaceX days, it would have been just the lack of understanding of what it takes to develop a rocket. In that case I was off by, say, 200 percent. I think future programs might be off by anywhere from like 25 percent to 50 percent as opposed to 200 percent.
>
> So, I think generally you do want to have a timeline where, based on everything you know about, the schedule should be X,

and you execute towards that, but with the understanding that there will be all sorts of things that you don't know about that you will encounter that will push the date beyond that. It doesn't mean that you shouldn't have tried to aim for that date from the beginning because aiming for something else would have been an arbitrary time increase.

It's different to say, "Well, what do you promise people?" Because you want to try to promise people something that includes schedule margin. But in order to achieve the external promised schedule, you've got to have an internal schedule that's more aggressive than that. Sometimes you still miss the external schedule.

SpaceX, by the way, is not alone here. Being late is par for the course in the aerospace industry. It's not a question of if it's late, it's how late will the program be. I don't think an aerospace program has been completed on time since bloody World War II.

Dealing with the epically aggressive schedules and Musk's expectations has required SpaceX's engineers to develop a variety of survival techniques. Musk often asks for highly detailed proposals for how projects will be accomplished. The employees have learned never to break the time needed to accomplish something down into months or weeks. Musk wants day-by-day and hour-by-hour forecasts and sometimes even minute-by-minute countdowns, and the fallout from missed schedules is severe. "You had to put in when you would go to the bathroom," Brogan said. "I'm like, 'Elon, sometimes people need to take a long dump.'" SpaceX's top managers work together to, in essence, create fake schedules that they know will please Musk but that are basically impossible to achieve. This would not be such a horrible situation if the targets were kept internal. Musk, however, tends to quote these fake schedules to customers, unintentionally giving them false hope. Typically, it falls to Gwynne Shotwell, SpaceX's president, to clean up the resulting mess. She will either need to ring up a customer to give them a more realistic timeline or concoct a litany of excuses to explain away the inevitable delays. "Poor Gwynne," Brogan said. "Just to hear her on the phone with the customers is agonizing."

There can be no question that Musk has mastered the art of getting the most out of his employees. Interview three dozen SpaceX engineers and each one of them will have picked up on a managerial nuance that Musk has used to get people to meet his deadlines. One example from Brogan: Where a typical manager may set the deadline for the employee, Musk guides his engineers into taking ownership of their own delivery dates. "He doesn't say, 'You have to do this by Friday at two P.M.,'" Brogan said. "He says, 'I need the impossible done by Friday at two P.M. Can you do it?' Then, when you say yes, you are not working hard because he told you to. You're working hard for yourself. It's a distinction you can feel. You have signed up to do your own work." And by recruiting hundreds of bright, self-motivated people, SpaceX has maximized the power of the individual. One person putting in a sixteen-hour day ends up being much more effective than two people working eight-hour days together. The individual doesn't have to hold meetings, reach a consensus, or bring other people up to speed on a project. He just keeps working and working and working. The ideal SpaceX employee is someone like Steve Davis, the director of advanced projects at SpaceX. "He's been working sixteen hours a day every day for years," Brogan said. "He gets more done than eleven people working together."

To find Davis, Musk called a teaching assistant* in Stanford's aeronautics department and asked him if there were any

---

* Michael Colonno.

hardworking, bright master's and doctoral candidates who didn't have families. The TA pointed Musk to Davis, who was pursuing a master's degree in aerospace engineering to add to degrees in finance, mechanical engineering, and particle physics. Musk called Davis on a Wednesday and offered him a job the following Friday. Davis was the twenty-second SpaceX hire and has ended up the twelfth most senior person still at the company. He turned thirty-five in 2014.

Davis did his tour of duty on Kwaj and considered it the greatest time of his life. "Every night, you could either sleep by the rocket in this tent shelter where the geckos crawled all over you or take this one-hour boat ride that made you seasick back to the main island," he said. "Every night, you had to pick the pain that you remembered least. You got so hot and exhausted. It was just amazing." After working on the Falcon 1, Davis moved to the Falcon 9 and then Dragon.

The Dragon capsule took SpaceX four years to design. It's likely the fastest project of its ilk done in the history of the aerospace industry. The project started with Musk and a handful of engineers, most of them under thirty years old, and peaked at one hundred people.* They cribbed from past capsule work and read over every paper published by NASA and other aeronautics bodies around projects like Gemini and Apollo. "If you go search for something like Apollo's reentry guidance algorithm, there are these great databases that will just spit out the answer," Davis said. The engineers at SpaceX then had to figure out how

---

*   According to Musk, "The early Dragon Version 1 work was just me and maybe three or four engineers, as we were living hand to mouth and had no idea if NASA would award us a contract. Technically, there was Magic Dragon before that, which was much simpler, as it had no NASA requirements. Magic Dragon was just me and some high altitude balloon guys in the U.K."

to advance these past efforts and bring the capsule into the modern age. Some of the areas of improvement were obvious and easily accomplished, while others required more ingenuity. Saturn 5 and Apollo had colossal computing bays that produced only a fraction of the computer horsepower that can be achieved today on, say, an iPad. The SpaceX engineers knew they could save a lot of room by cutting out some of the computers while also adding capabilities with their more powerful equipment. The engineers decided that while Dragon would look a lot like Apollo, it would have steeper wall angles, to clear space for gear and for the astronauts that the company hoped to fly. SpaceX also got the recipe for its heat shield material, called PICA, through a deal with NASA. The SpaceX engineers found out how to make the PICA material less expensively and improved the underlying recipe so that Dragon—from day one—could withstand the heat of a reentry coming back from Mars.* The total cost for Dragon came in at $300 million, which would be on the order of 10 to 30 times less than capsule projects built by other companies. "The metal comes in, we roll it out, weld it, and make things," Davis said. "We build almost everything in-house. That is why the costs have come down."

Davis, like Brogan and plenty of other SpaceX engineers, has had Musk ask for the seemingly impossible. His favorite request dates back to 2004. SpaceX needed an actuator that would trigger the gimbal action used to steer the upper stage of Falcon 1. Davis had never built a piece of hardware before in his life and natu-

---

* NASA researchers studying the Dragon design have noticed several features of the capsule that appear to have been purpose built from the get-go to accommodate a landing on Mars. They've published a couple of papers explaining how it could be feasible for NASA to fund a mission to Mars in which a Dragon capsule picks up samples and returns them to Earth.

rally went out to find some suppliers who could make an electro-mechanical actuator for him. He got a quote back for $120,000. "Elon laughed," Davis said. "He said, 'That part is no more complicated than a garage door opener. Your budget is five thousand dollars. Go make it work.'" Davis spent nine months building the actuator. At the end of the process, he toiled for three hours writing an e-mail to Musk covering the pros and cons of the device. The e-mail went into gory detail about how Davis had designed the part, why he had made various choices, and what its cost would be. As he pressed send, Davis felt anxiety surge through his body knowing that he'd given his all for almost a year to do something an engineer at another aerospace company would not even attempt. Musk rewarded all of this toil and angst with one of his standard responses. He wrote back, "Ok." The actuator Davis designed ended up costing $3,900 and flew with Falcon 1 into space. "I put every ounce of intellectual capital I had into that e-mail and one minute later got that simple response," Davis said. "Everyone in the company was having that same experience. One of my favorite things about Elon is his ability to make enormous decisions very quickly. That is still how it works today."

Kevin Watson can attest to that. He arrived at SpaceX in 2008 after spending twenty-four years at NASA's Jet Propulsion Laboratory. Watson worked on a wide variety of projects at JPL, including building and testing computing systems that could withstand the harsh conditions of space. JPL would typically buy expensive, specially toughened computers, and this frustrated Watson. He daydreamed about ways to handcraft much cheaper, equally effective computers. While having his job interview with Musk, Watson learned that SpaceX needed just this type of thinking. Musk wanted the bulk of a rocket's computing systems to cost no more than $10,000. It was an insane figure by aerospace industry standards, where the avionics systems for a rocket typically cost well over $10 million. "In traditional aerospace, it would cost you more than ten thousand dollars just for the food at a meeting to discuss the cost of the avionics," Watson said.

During the job interview, Watson promised Musk that he could do the improbable and deliver the $10,000 avionics system. He began working on making the computers for Dragon right after being hired. The first system was called CUCU, pronounced "cuckoo." This communications box would go inside the International Space Station and communicate back with Dragon. A number of people at NASA referred to the SpaceX engineers as "the guys in the garage" and were cynical about the start-up's ability to do much of anything, including building this type of machine. But SpaceX produced the communication computer in record time, and it ended up as the first system of its kind to pass NASA's protocol tests on the first try. NASA officials were forced to say "cuckoo" over and over again during meetings—a small act of defiance SpaceX had planned all along to torture NASA. As the months went on, Watson and other engineers built out the complete computing systems for Dragon and then adapted the technology for Falcon 9. The result was a fully redundant avionics platform that used a mix of off-the-shelf computing gear and products built in-house by SpaceX. It cost a bit more than $10,000 but came close to meeting Musk's goal.

SpaceX reinvigorated Watson, who had become disenchanted with JPL's acceptance of wasteful spending and bureaucracy. Musk had to sign off on every expenditure over $10,000. "It was his money that we were spending, and he was keeping an eye on it, as he damn well should," Watson said. "He made sure nothing stupid was happening." Decisions were made quickly during weekly meetings, and the entire company bought into them. "It was amazing how fast people would adapt to what came out of those meetings," Watson said. "The entire ship could turn ninety degrees instantly. Lockheed Martin could never do anything like that." Watson continued:

> *Elon is brilliant. He's involved in just about everything. He understands everything. If he asks you a question, you learn very quickly not to go give him a gut reaction. He wants answers that get down to the fundamental laws of physics. One thing he understands really well is the physics of the rockets. He understands that like nobody else. The stuff I have seen him do in his head is crazy. He can get in discussions about flying a satellite and whether we can make the right orbit and deliver Dragon at the same time and solve all these equations in real time. It's amazing to watch the amount of knowledge he has accumulated over the years. I don't want to be the person who ever has to compete with Elon. You might as well leave the business and find something else fun to do. He will outmaneuver you, outthink you, and out-execute you.*

One of Watson's top discoveries at SpaceX was the test bed on the third floor of the Hawthorne factory. SpaceX has test versions of all the hardware and electronics that go into a rocket laid out on metal tables. It has in effect replicated the innards of a rocket end to end in order to run thousands of flight simulations. Someone "launches" the rocket from a computer and then every piece of mechanical and computing hardware is monitored with sensors. An engineer can tell a valve to open, then check to see if it opened, how quickly it opened, and the level of current running to it. This testing apparatus lets SpaceX engineers practice ahead of launches and figure out how they would deal with all manner of anomalies. During the actual flights, SpaceX has people in the test facility who can replicate errors seen on Falcon or Dragon and make adjustments accordingly. SpaceX has made numerous changes on the fly with this system. In one case someone spotted an error in a software file in the hours right before a launch. SpaceX's engineers changed the file, checked how it affected the test hardware, and, when no problems were detected, sent the file to the Falcon 9, waiting on the launchpad, all in less than thirty minutes. "NASA wasn't used to this," Watson said. "If something went wrong with the shuttle, everyone was just resigned to waiting three weeks before they could try and launch again."[12]

From time to time, Musk will send out an e-mail to the entire company to enforce a new policy or let them know about something that's bothering him. One of the more famous e-mails arrived in May 2010 with the subject line: Acronyms Seriously Suck:

> There is a creeping tendency to use made up acronyms
> at SpaceX. Excessive use of made up acronyms is a
> significant impediment to communication and keeping
> communication good as we grow is incredibly important.
> Individually, a few acronyms here and there may not seem
> so bad, but if a thousand people are making these up, over
> time the result will be a huge glossary that we have to
> issue to new employees. No one can actually remember
> all these acronyms and people don't want to seem dumb
> in a meeting, so they just sit there in ignorance. This is
> particularly tough on new employees.
>
> That needs to stop immediately or I will take drastic
> action—I have given enough warnings over the years.
> Unless an acronym is approved by me, it should not enter
> the SpaceX glossary. If there is an existing acronym that
> cannot reasonably be justified, it should be eliminated, as I
> have requested in the past.
>
> For example, there should be no "HTS" [horizontal
> test stand] or "VTS" [vertical test stand] designations for

test stands. Those are particularly dumb, as they contain unnecessary words. A "stand" at our test site is obviously a *test* stand. VTS-3 is four syllables compared with "Tripod," which is two, so the bloody acronym version actually takes longer to say than the name!

The key test for an acronym is to ask whether it helps or hurts communication. An acronym that most engineers outside of SpaceX already know, such as GUI, is fine to use. It is also ok to make up a few acronyms/contractions every now and again, assuming I have approved them, eg MVac and M9 instead of Merlin 1C-Vacuum or Merlin 1C-Sea Level, but those need to be kept to a minimum.

This was classic Musk. The e-mail is rough in its tone and yet not really unwarranted for a guy who just wants things done as efficiently as possible. It obsesses over something that other people might find trivial and yet he has a definite point. It's comical in that Musk wants all acronym approvals to run directly through him, but that's entirely in keeping with the hands-on management style that has, mainly, worked well at both SpaceX and Tesla. Employees have since dubbed the acronym policy the ASS Rule.

The guiding principle at SpaceX is to embrace your work and get stuff done. People who await guidance or detailed instructions languish. The same goes for workers who crave feedback. And the absolute worst thing that someone can do is inform Musk that what he's asking is impossible. An employee could be telling Musk that there's no way to get the cost on something like that actuator down to where he wants it or that there is simply not enough time to build a part by Musk's deadline. "Elon will say, 'Fine. You're off the project, and I am now the CEO of the project. I will do your job and be CEO of two companies at the same time. I will deliver it,'" Brogan said. "What's crazy is that Elon actually does it. Every time he's fired someone and taken their job, he's delivered on whatever the project was."

It is jarring for both parties when the SpaceX culture rubs against more bureaucratic bodies like NASA, the U.S. Air Force, and the Federal Aviation Administration. The first inklings of these difficulties appeared on Kwaj, where government officials sometimes questioned what they saw as SpaceX's cavalier approach to the launch process. There were times when SpaceX would want to make a change to its launch procedures and any such change would require a pile of paperwork. SpaceX, for example, would have written down all the steps needed to replace a filter—put on gloves, wear safety goggles, remove a nut—and then want to alter this procedure or use a different type of filter. The FAA would need a week to review the new process before SpaceX could actually go about changing the filter on the rocket, a lag that both the engineers and Musk found ridiculous. On one occasion after this type of thing happened, Musk laid into an FAA official while on a conference call with members of the SpaceX team and NASA. "It got hot and heated, and he berated this guy on a personal level for like ten minutes," Brogan said.

Musk did not recall this incident but did remember other confrontations with the FAA. One time he compiled a list of things an FAA subordinate had said during a meeting that Musk found silly and sent the list along to the guy's boss. "And then his dingbat manager sent me this long e-mail about how he had been in the shuttle program and in charge of twenty launches or something like that and how dare I say that the other guy was wrong," Musk said. "I told him, 'Not only is he wrong, and let me rearticulate the reasons, but you're wrong, and let me articulate the reasons.' I don't think he sent me another e-mail after that. We're trying to have a really big impact on the space industry. If the rules are such that you can't make progress, then you have to fight the rules.

"There is a fundamental problem with regulators. If a regulator agrees to change a rule and something bad happens, they could easily lose their career. Whereas if they change a rule and something good happens, they don't even get a reward. So, it's very asymmetric. It's then very easy to understand why regulators resist changing the rules. It's because there's a big punishment on one side and no reward on the other. How would any rational person behave in such a scenario?"

In the middle of 2009, SpaceX hired Ken Bowersox, a former astronaut, as its vice president of astronaut safety and mission assurance. Bowersox fit the mold of recruit prized by a classic big aerospace company. He had a degree in aerospace engineering from the U.S. Naval Academy, had been a test pilot in the air force, and flew on the space shuttle a handful of times. Many people within SpaceX saw his arrival at the company as a good thing. He was considered a diligent, dignified sort who would provide a second set of eyes to many of SpaceX's procedures, checking to make sure the company went about things in a safe, standardized manner. Bowersox ended up smack in the middle of the constant pull and push at SpaceX between doing things efficiently and agonizing over traditional procedures. He and Musk were increasingly at odds as the months passed, and Bowersox started to feel as if his opinions were being ignored. During one incident in particular, a part made it all the way to the test stand with a major flaw—described by one engineer as the equivalent of a coffee cup not having a bottom—instead of being caught at the factory. According to observers, Bowersox argued that SpaceX should go back and investigate the process that led to the mistake and fix its root cause. Musk had already decided that he knew the basis of the problem and dismissed Bowersox after a couple of years on the job. (Bowersox declined to speak on the record about his time at SpaceX.) A number of people inside SpaceX saw the Bowersox incident as an example of Musk's hard-charging manner undermining some much-needed process. Musk had a totally different take on the situation, casting Bowersox as not being up to the engineering demands at SpaceX.

A handful of high-ranking government officials gave me their candid takes on Musk, albeit without being willing to put their names to the remarks. One found Musk's treatment of air force generals and military men of similar rank appalling. Musk has been known to let even high-ranking officials have it when he thinks they're off base and is not apologetic about this. Another could not believe it when Musk would call very intelligent people idiots. "Imagine the worst possible way that could come out, and it would come out," this person said. "Life with Elon is like being in a very intimate married couple. He can be so gentle and loyal and then really hard on people when it isn't necessary." One former official felt that Musk would need to temper himself better in the years to come if SpaceX was to keep currying favor with the military and government agencies in its bid to defeat the incumbent contractors. "His biggest enemy will be himself and the way he treats people," this person said.

When Musk rubs outsiders the wrong way, Shotwell is often there to try to smooth over the situation. Like Musk, she has a salty tongue and a fiery personality, but Shotwell is willing to play the role of the conciliator. These skills have allowed her to handle the day-to-day operations at SpaceX, leaving Musk to focus on the company's overall strategy, the product designs, marketing, and motivating employees. Like all of Musk's most trusted lieutenants, Shotwell has been willing to stay largely in the background, do her work, and focus on the company's cause.

Shotwell grew up in the suburbs of Chicago, the daughter of

an artist (mom) and a neurosurgeon (dad). She played the part of a bright, pretty girl, getting straight A's at school and joining the cheerleading squad. Shotwell had not expressed a major inclination toward the sciences and knew only one version of an engineer—the guy who drives a train. But there were clues that she was wired a bit different. She was the daughter who mowed the lawn and helped put the family basketball hoop together. In third grade, Shotwell developed a brief interest in car engines, and her mom bought a book detailing how they work. Later, in high school, Shotwell's mom forced her to attend a lecture at the Illinois Institute of Technology on a Saturday afternoon. As Shotwell listened to one of the panels, she grew enamored with a fifty-year-old mechanical engineer. "She had these beautiful clothes, this suit and shoes that I loved," Shotwell said. "She was tall and carried off the heels really well." Shotwell chatted with the engineer after the talk, learning about her job. "That was the day I decided to become a mechanical engineer," she said.

Shotwell went on to receive an undergraduate degree in mechanical engineering and a master's degree in applied mathematics from Northwestern University. Then she took a job at Chrysler. It was a type of management training program meant for hotshot recent graduates who appeared to have leadership potential. Shotwell started out going to auto mechanics school—"I loved that"—and then from department to department. While working on engines research, Shotwell found that there were two very expensive Cray supercomputers sitting idle because none of the veterans knew how to use them. A short while later, she logged onto the computers and set them up to run computational fluid dynamics, or CFD, operations to simulate the performance of valves and other components. The work kept Shotwell interested, but the environment started to grate on her. There were rules for everything, including lots of union regulations around who could operate certain machines. "I picked up a tool once, and got written up," she said. "Then I opened a bottle of liquid nitrogen and got written up. I started thinking that the job was not what I had anticipated it would be."

Shotwell pulled out of the Chrysler training program, regrouped at home, and then briefly pursued her doctorate in applied mathematics. While back on the Northwestern campus, one of her professors mentioned an opportunity at the Aerospace Corporation. Anything but a household name, Aerospace Corporation has been headquartered in El Segundo since 1960, serving as a kind of neutral, nonprofit organization that advises the air force, NASA, and other federal bodies on space programs. The company has a bureaucratic feel but has proved very useful over the years with its research activities and ability to champion and nix costly endeavors. Shotwell started at Aerospace in October 1988 and worked on a wide range of projects. One job required her to develop a thermal model that depicted how temperature fluctuations in the space shuttle's cargo bay affected the performance of equipment on various payloads. She spent ten years at Aerospace and honed her skills as a systems engineer. By the end, though, Shotwell had become irritated by the pace of the industry. "I didn't understand why it had to take fifteen years to make a military satellite," she said. "You could see my interest was waning."

For the next four years, Shotwell worked at Microcosm, a space start-up just down the road from the Aerospace Corporation, and became the head of its space systems division and business development. Boasting a combination of smarts, confidence, direct talk, and good looks, Shotwell developed a reputation as a strong saleswoman. In 2002, one of her coworkers, Hans Koenigsmann, left for SpaceX. Shotwell took Koenigsmann out for a going-away lunch and dropped him off at SpaceX's then rinky-dink headquarters. "Hans told me to go in and meet Elon," Shotwell said. "I did, and that's when I told him, 'You need a good business development person.'" The next day Mary Beth Brown called Shotwell and told her that Musk wanted to interview her for the new vice president of business development position. Shotwell ended up as employee No. 7. "I gave three weeks' notice at Microcosm and remodeled my bathroom because I knew I would not have a life after taking the job," she said.

Through the early years of SpaceX, Shotwell pulled off the miraculous feat of selling something the company did not have. It took SpaceX so much longer than it had planned to have a successful flight. The failures along the way were embarrassing and bad for business. Nonetheless, Shotwell managed to sell about a dozen flights to a mix of government and commercial customers before SpaceX put its first Falcon 1 into orbit. Her deal-making skills extended to negotiating the big-ticket contracts with NASA that kept SpaceX alive during its leanest years, including a $278 million contract in August 2006 to begin work on vehicles that could ferry supplies to the ISS. Shotwell's track record of success turned her into Musk's ultimate confidante at SpaceX, and at the end of 2008, she became president and chief operating officer at the company.

Part of Shotwell's duties include reinforcing the SpaceX culture as the company grows larger and larger and starts to resemble the traditional aerospace giants that it likes to mock. Shotwell can switch on an easygoing, affable air and address the entire company during a meeting or convince a collection of possible recruits why they should sign up to be worked to the bone. During one such meeting with a group of interns, Shotwell pulled about a hundred people into the corner of the cafeteria. She wore high-heel black boots, skintight jeans, a tan jacket, and a scarf and had big hoop earrings dangling beside her shoulder-length blond hair. Pacing back and forth in front of the group with a microphone in hand, she asked them to announce what school they came from and what project they were working on while at SpaceX. One student went to Cornell and worked on Dragon, another went to USC and did propulsion system design, and another went to the University of Illinois and worked with the aerodynamics group. It took about thirty minutes to make it all the way around the room, and the students were, at least by academic pedigree and bright-eyed enthusiasm, among the most impressive youngsters in the world. The students peppered Shotwell with questions—her best moment, her advice for being successful, SpaceX's competitive threats—and she replied with a mix of earnest answers and rah-rah stuff. Shotwell made sure to emphasize the lean, innovative edge SpaceX has over the more traditional aerospace companies. "Our competitors are scared shitless of us," Shotwell told the group. "The behemoths are going to have to figure out how to get it together and compete. And it is our job to have them die."

One of SpaceX's biggest goals, Shotwell said, was to fly as often as possible. The company has never sought to make a fortune off each flight. It would rather make a little on each launch and keep the flights flowing. A Falcon 9 flight costs $60 million, and the company would like to see that figure drop to about $20 million through economies of scale and improvements in launch technology. SpaceX spent $2.5 billion to get four Dragon capsules to the ISS, nine flights with the Falcon 9, and five flights with the Falcon 1. It's a price-per-launch total that the rest of the players in the industry cannot comprehend let alone aspire to. "I don't know what those guys do with their money," Shotwell said. "They are smoking it. I just don't know." As Shotwell saw it, a number of new nations were showing interest in launches, eyeing communications technology as essential to growing their economies and leveling their status with developed nations. Cheaper flights would help SpaceX take the majority of the business from that new customer set. The company also expected to participate in an expanding market for human flights. SpaceX has never had any interest in doing the five-minute tourist flights to low Earth orbit like Virgin Galactic and XCOR. It does, however, have the ability to carry researchers to orbiting habitats being built by Bigelow Aerospace and to orbiting science labs being constructed by various countries. SpaceX will also start making its own satellites, turning the company into a one-stop space shop. All of these plans hinge on SpaceX being able to prove that it can fly on schedule every month and churn through the $5 billion backlog of launches. "Most of our customers signed up early and wanted to be supportive and got good deals on their missions," she said. "We are in a phase now where we need to launch on time and make launching Dragons more efficient."

For a short while, the conversation with the interns bogged down. It turned to some of the annoyances of SpaceX's campus. The company leases its facility and has not been able to build things like a massive parking structure that would make life easier for its three-thousand-person workforce. Shotwell promised that more parking, more bathrooms, and more of the freebies that technology start-ups in Silicon Valley offer their employees would be on the way. "I want a day care," she said.

But it was while discussing SpaceX's grandest missions that Shotwell really came into her own and seemed to inspire the interns. Some of them clearly dreamed of becoming astronauts, and Shotwell said that working at SpaceX was almost certainly their best chance to get to space now that NASA's astronaut corps had dwindled. Musk had made designing cool-looking, "non–Stay Puft" spacesuits a personal priority. "They can't be clunky and nasty," Shotwell said. "You have to do better than that." As for where the astronauts would go: well, there were the space habitats, the moon, and, of course, Mars as options. SpaceX has already started testing a giant rocket, called the Falcon Heavy, that will take it much farther into space than the Falcon 9, and it has another, even larger spaceship on the way. "Our Falcon Heavy rocket will not take a busload of people to Mars," she said. "So, there's something after Heavy. We're working on it." To make something like that vehicle happen, she said, the SpaceX employees needed to be effective and pushy. "Make sure your output is high," Shotwell said. "If we're throwing a bunch of shit in your way, you need to be mouthy about it. That's not a quality that's widely accepted elsewhere, but it is at SpaceX." And, if that sounded harsh, so be it. As Shotwell saw it, the commercial space race was coming down to SpaceX and China and that's it. And in the bigger picture, the race was on to ensure man's survival. "If you hate people and think human extinction is okay, then fuck it," Shotwell said. "Don't go to space. If you think it is worth humans doing some risk management and finding a second place to go live, then you should be focused on this issue and willing to spend some money. I am pretty sure we will be selected by NASA to drop landers and rovers off on Mars. Then the first SpaceX mission will be to drop off a bunch of supplies, so that once people get there, there will be places to live and food to eat and stuff for them to do."

It's talk like this that thrills and amazes people in the aerospace industry, who have long been hoping that some company would come along and truly revolutionize space travel. Aeronautics experts will point out that twenty years after the Wright brothers started their experiments, air travel had become routine. The launch business, by contrast, appears to have frozen. We've been to the moon, sent research vehicles to Mars, and explored the solar system, but all of these things are still immensely expensive one-off projects. "The cost remains extraordinarily high because of the rocket equation," said Carol Stoker, the planetary scientist at NASA. Thanks to military and government contracts from agencies like NASA, the aerospace industry has historically had massive budgets to work with and tried to make the biggest, most reliable machines it could. The business has been tuned to strive for maximum performance, so that the aerospace contractors can say they met their requirements. That strategy makes sense if you're trying to send up a $1 billion military satellite for the U.S. government and simply cannot afford for the payload to blow up. But on the whole, this approach stifles the pursuit of other endeavors. It leads to bloat and excess and a crippling of the commercial space industry.

Outside of SpaceX, the American launch providers are no longer competitive against their peers in other countries. They have limited launch abilities and questionable ambition. SpaceX's main competitor for domestic military satellites and other large payloads is United Launch Alliance (ULA), a joint venture formed in 2006 when Boeing and Lockheed Martin combined forces. The thinking at the time about the union was that the government did not have enough business for two companies and that combining the research and manufacturing work of Boeing and Lockheed would result in cheaper, safer launches. ULA has leaned on decades of work around the Delta (Boeing) and Atlas (Lockheed) launch vehicles and has flown many dozens of rockets successfully, making it a model of reliability. But neither the joint venture nor Boeing nor Lockheed, both of which can offer commercial services on their own, come close to competing on price against SpaceX, the Russians, or the Chinese. "For the most part, the global commercial market is dominated by Arianespace [Europe], Long March [China] or Russian vehicles," said Dave Bearden, the general manager of civil and commercial programs at the Aerospace Corporation. "There are just different labor rates and differences in the way they are built."

To put things more bluntly, ULA has turned into an embarrassment for the United States. In March 2014, ULA's then CEO, Michael Gass, faced off against Musk during a congressional hearing that dealt, in part, with SpaceX's request to take on more of the government's annual launch load. A series of slides were rolled out that showed how the government payments for launches have skyrocketed since Boeing and Lockheed went from a duopoly to a monopoly. According to Musk's math presented at the hearing, ULA charged $380 million per flight, while SpaceX would charge $90 million per flight. (The $90 million figure was higher than SpaceX's standard $60 million because the government has certain additional requirements for particularly sensitive launches.) By simply picking SpaceX as its launch provider, Musk pointed out, the government would save enough money to pay for the satellite going on the rocket. Gass had no real retort. He claimed Musk's figures for the ULA launch price were inaccurate but failed to provide a figure of his own. The hearing also came as tensions between the United States and Russia were running high due to Russia's aggressive actions in Ukraine. Musk rightly noted that the United States could soon be placing sanctions on Russia that could carry over to aerospace equipment. ULA, as it happens, relies on Russian-made engines to send up sensitive U.S. military equipment in its Atlas V rockets. "Our Falcon 9 and Falcon Heavy launch vehicles are truly American," Musk said. "We design and manufacture our rockets in California and Texas." Gass countered that ULA had bought a two-year supply of Russian engines and purchased the blueprints to the machines and had them translated from Russian to English, and he said this with a straight face. (A few months after the hearing, ULA replaced Gass as CEO and signed a deal with Blue Origin to develop American-made rockets.)

Some of the most disheartening moments of the hearing arrived when Senator Richard Shelby of Alabama took the microphone for questioning. ULA has manufacturing facilities in Alabama and close ties to the senator. Shelby felt compelled to play the role of hometown booster by repeatedly pointing out that ULA had enjoyed sixty-eight successful launches and then asking Musk what he made of that accomplishment. The aerospace industry stands as one of Shelby's biggest donors and he's ended up surprisingly pro-bureaucracy and anticompetition when it comes to getting things into space. "Typically competition results in better quality and lower-priced contracts—but the launch market is not typical," Shelby said. "It is limited demand framed by government-industrial policies." The March hearing in which Shelby made these statements would turn out to be something of a sham. The government had agreed to put fourteen of its sensitive launches up for bid instead of just awarding them directly to ULA. Musk had come to Congress to present his case for why SpaceX made sense as a viable candidate for those and other launches. The day after the hearing, the air force cut the number of launches up for bid from fourteen to between seven and one. One month later, SpaceX filed a lawsuit against the air force asking for a chance to earn its launch business. "SpaceX is not seeking to be awarded contracts for these launches," the company said on its freedomtolaunch.com website. "We are simply seeking the right to compete."*

---

* The politicking in the space business can get quite nasty. Lori Garver, the former deputy administrator of NASA, spent years fighting to open up NASA contracts so that private companies could bid on things like resupplying the ISS. Her position of fostering a strong relationship between NASA and the private sector won out in the end but at a cost. "I had death threats and fake anthrax sent to me," she said. Garver also ran across SpaceX competitors that tried to spread unfounded gossip about the company and Musk. "They claimed he was in violation of tax laws in South Africa and had another, secret family there. I said, 'You're making this stuff up.' We're lucky that people with such long-term visions as Elon, Jeff Bezos, and Robert

SpaceX's main competitor for ISS resupply missions and commercial satellites in the United States is Orbital Sciences Corporation. Founded in Virginia in 1982, the company started out not unlike SpaceX, as the new kid that raised outside funding and focused on putting smaller satellites into low-Earth orbit. Orbital is more experienced, although it has a limited roster of machine types. Orbital depends on suppliers, including Russian and Ukrainian companies, for its engines and rocket bodies, making it more of an assembler of spacecraft than a true builder like SpaceX. And, also unlike SpaceX, Orbital's capsules cannot withstand the journey back from the ISS to Earth, so it's unable to return experiments and other goods. In October 2014, one of Orbital's rockets blew up on the launchpad. With its ability to launch on hold while it investigated the incident, Orbital reached out to SpaceX for help. It wanted to see if Musk had any extra capacity to take care of some of Orbital's customers. The company also signaled that it would move away from using Russian engines as well.

As for getting humans to space, SpaceX and Boeing were the victors in a four-year NASA competition to fly astronauts to the ISS. SpaceX will get $2.6 billion, and Boeing will get $4.2 billion to develop their capsules and ferry people to the ISS by 2017. The companies would, in effect, be replacing the space shuttle and restoring the United States' ability to conduct manned flights. "I actually don't mind that Boeing gets twice as much money for meeting the same NASA requirements as SpaceX with worse technology," Musk said. "Having two companies involved is better for the advancement of human spaceflight."

SpaceX had once looked like it too would be a one-trick pony.

Bigelow [founder of the aerospace company that bears his name] got rich. It's nuts that people would want to vilify Elon. He might say some things that rub people the wrong way, but, at some point, the being nice to everyone thing doesn't work."

The company's original plans were to have the smallish Falcon 1 function as its primary workhorse. At $6 million to $12 million per flight, the Falcon 1 was by far the cheapest means of getting something into orbit, thrilling people in the space industry. When Google announced its Lunar X Prize in 2007—$30 million in awards to people who could land a robot on the moon—many of the proposals that followed selected the Falcon 1 as their preferred launch vehicle because it seemed like the only reasonably priced option for getting something to the moon. Scientists around the world were equally excited, thinking that for the first time they had a means of placing experiments into orbit in a cost-effective way. But for all the enthusiastic talk about the Falcon 1, the demand never arrived. "It became very clear that there was a huge need for the Falcon 1 but no money for it," said Shotwell. "The market has to be able to sustain a certain amount of vehicles, and three Falcon 1s per year does not make a business." The last Falcon 1 launch took place in July 2009 from Kwajalein, when SpaceX carried a satellite into orbit for the Malaysian government. People in the aerospace industry have been grumbling ever since. "We gave Falcon 1 a hell of a shot," Shotwell said. "I was emotional about it and disappointed. I'd anticipated a flood of orders but, after eight years, they just did not come."

SpaceX has since expanded its launch capabilities at a remarkable pace and looks like it might be on the verge of getting that $12 million per flight option back. In June 2010, the Falcon 9 flew for the first time and orbited Earth successfully. In December 2010, SpaceX proved that the Falcon 9 could carry the Dragon capsule into space and that the capsule could be recovered safely after an ocean landing.* It became the first commercial company ever to

---

* On this flight, SpaceX secretly placed a wheel of cheese inside the Dragon capsule. It was the same one Jeff Skoll had given Musk back in the mice-to-Mars days.

pull off this feat. Then, in May 2012, SpaceX went through the most significant moment in the company's history since that first successful launch on Kwajalein.

On May 22, at 3:44 A.M., a Falcon 9 rocket took off from the Kennedy Space Center in Cape Canaveral, Florida. The rocket did its yeoman-like work boosting Dragon into space. Then the capsule's solar panels fanned out and Dragon became dependent on its eighteen Draco thrusters, or small rocket engines, to guide its path to the International Space Station. The SpaceX engineers worked in shifts—some of them sleeping on cots at the factory—as it took the capsule three days for Dragon to make its journey. They spent most of the time observing Dragon's flight and checking to see that its sensor systems were picking up the ISS. Originally, Dragon planned to dock with the ISS around 4 A.M. on the twenty-fifth, but as the capsule approached the space station, an unexpected glint kept throwing off the calculations of a laser used to measure the distance between Dragon and the ISS. "I remember it being two and a half hours of struggle," Shotwell said. Her outfit of Uggs, a fishnet sweater, and leggings started to feel like pajamas as the night wore on, and the engineers battled this unplanned difficulty. Fearing all the time that the mission would be aborted, SpaceX decided to upload some new software to the Dragon that would cut the size of the visual frame used by the sensors to eliminate the effect of the sunlight on the machine. Then, just before 7 A.M., Dragon got close enough to the ISS for Don Pettit, an astronaut, to use a fifty-eight-foot robotic arm to reach out and grab the resupply capsule. "Houston, Station, it looks like we've got us a dragon by the tail," Pettit said.[13]

"I'd been digesting my guts," Shotwell said. "And then I am drinking champagne at six in the morning." About thirty people were in the control room when the docking happened. Over the next couple of hours, workers streamed into the SpaceX factory to soak up the elation of the moment. SpaceX had set another first, as the only private company to dock with the ISS. A couple of months later SpaceX received $440 million from NASA to keep developing Dragon so that it could transport people. "Elon is changing the way aerospace business is done," said NASA's Stoker. "He's managed to keep the safety factor up while cutting costs. He's just taken the best things from the tech industry like the open-floor office plans and having everyone talking and all this human interaction. It's a very different way to most of the aerospace industry, which is designed to produce requirements documents and project reviews."

In May 2014, Musk invited the press to SpaceX's headquarters to demonstrate what some of that NASA money had bought. He unveiled the Dragon V2, or version two, spacecraft. Unlike most executives, who like to show their products off at trade shows or daytime events, Musk prefers to hold true Hollywood-style galas in the evenings. People arrived in Hawthorne by the hundreds and snacked on hors d'oeuvres until the 7:30 P.M. showing. Musk appeared wearing a purplish velvet jacket and popping open the capsule's door with a bump of his fist like the Fonz. What he revealed was spectacular. The cramped quarters of past capsules were gone. There were seven thin, sturdy, contoured seats arranged with four seats close to the main console and a row of three seats in the back. Musk walked around in the capsule to show how roomy it was and then plopped down in the central captain's chair. He reached up and unlocked a four-paneled flat-screen console that gracefully slid down right in front of the first row of seats. In the middle of the console was a joystick for fly-

---

* Musk explained the look to me in a way that only he can. "I went for a similar style to the Model S (it uses the same screens as Model S upgraded for space ops), but kept the aluminum isogrid uncovered for a more exotic feel."

ing the aircraft and some physical buttons for essential functions that astronauts could press in case of an emergency or a malfunctioning touch-screen. The inside of the capsule had a bright, metallic finish. Someone had finally built a spaceship worthy of scientist and moviemaker dreams.

There was substance to go with the style. The Dragon 2 will be able to dock with the ISS and other space habitats automatically without needing the intervention of a robotic arm. It will run on a SuperDraco engine—a thruster made by SpaceX and the first engine ever built completely by a 3-D printer to go into space. This means that a machine guided by a computer formed the engine out of single piece of metal—in this case the high-strength alloy Inconel—so that its strength and performance should exceed anything built by humans by welding various parts together. And most mind-boggling of all, Musk revealed that the Dragon 2 will be able to land anywhere on Earth that SpaceX wants by using the SuperDraco engines and thrusters to come to a gentle stop on the ground. No more landings at sea. No more throwing spaceships away. "That is how a twenty-first-century spaceship should land," Musk said. "You can just reload propellant and fly again. So long as we continue to throw away rockets and spacecraft, we will never have true access to space."

The Dragon 2 is just one of the machines that SpaceX continues to develop in parallel. One of the company's next milestones will be the first flight of the Falcon Heavy, which is designed to be the world's most powerful rocket.* SpaceX has found a way to

---

* Rather insanely, NASA is building a next-generation, giant spaceship that could one day get to Mars even though SpaceX is building the same type of craft—the Falcon Heavy—on its own. NASA's program is budgeted to cost $18 billion, although government studies say that figure is very conservative. "NASA has no fucking business doing this," said Andrew Beal, the billionaire investor and onetime commercial space entrepreneur. "The whole space shuttle system was a disas-

combine three Falcon 9s into a single craft with 27 of the Merlin engines and the ability to carry more than 53 metric tons of stuff into orbit. Part of the genius of Musk and Mueller's designs is that SpaceX can reuse the same engine in different configurations—from the Falcon 1 up to the Falcon Heavy—saving on cost and time. "We make our main combustion chambers, turbo pump, gas generators, injectors, and main valves," Mueller said. "We have complete control. We have our own test site, while most of the other guys use government test sites. The labor hours are cut in half and so is the work around the materials. Four years ago, we could make two rockets a year and now we can make twenty a year." SpaceX boasts that the Falcon Heavy can take up twice the payload of the nearest competitor—the Delta IV Heavy from Boeing/ULA—at one-third the cost. SpaceX is also busy building a spaceport from the ground up. The goal is to be able to launch many rockets an hour from this facility located in Brownsville, Texas, by automating the processes needed to stand a rocket up on the pad, fuel it, and send it off.

Just as it did in the early days, SpaceX continues to experiment with these new vehicles during actual launches in ways that other companies would dare not do. SpaceX will often announce that it's trying out a new engine or its landing legs and place the emphasis on that one upgrade in the marketing material leading

---

ter. They're fucking clueless. Who in their right mind would use huge solid boosters, especially ones built in segments requiring dynamic seals? They are so lucky they only had one disastrous failure of the boosters." Beal's firm criticisms come from years of watching the government compete against private space companies by subsidizing the construction of spacecraft and launches. His company Beal Aerospace quit the business because the government kept funding competing rockets. "Governments around the world have spent billions trying to do what Elon is doing, and they have failed," he said. "We have to have governments, but the idea that the government goes out and competes with companies is fucking nuts."

up to a launch. It's common, though, for SpaceX to test out a dozen other objectives in secret during a mission. Musk essentially asks employees to do the impossible on top of the impossible. One former SpaceX executive described the working atmosphere as a perpetual-motion machine that runs on a weird mix of dissatisfaction and eternal hope. "It's like he has everyone working on this car that is meant to get from Los Angeles to New York on one tank of gas," this executive said. "They will work on the car for a year and test all of its parts. Then, when they set off for New York after that year, all of the vice presidents think privately that the car will be lucky to get to Las Vegas. What ends up happening is that the car gets to New Mexico—twice as far as they ever expected—and Elon is still mad. He gets twice as much as anyone else out of people."

There's a degree to which it's just never enough for Musk, no matter what it is. Case in point: the December 2010 launch in which SpaceX got the Dragon capsule to orbit Earth and return successfully. This had been one of the company's great achievements, and people had worked tirelessly for months, if not years. The launch had taken place on December 8, and SpaceX had a Christmas party on December 16. About ninety minutes before the party started, Musk had called his top executives to SpaceX for a meeting. Six of them, including Mueller, were decked out in party attire and ready to celebrate the holidays and SpaceX's historic achievement around Dragon. Musk laid into them for about an hour because the truss structure for a future rocket was running behind schedule. "Their wives were sitting three cubes over waiting for the berating to end," Brogan said. Other examples of similar behavior have cropped up from time to time. Musk, for example, rewarded a group of thirty employees who had pulled off a tough project for NASA with bonuses that consisted of additional stock option grants. Many of the employees, seeking instant, more tangible gratification, demanded cash. "He chided us for not valuing the stock," Drew Eldeen, a former engineer, said. "He said, 'In the long run, this is worth a lot more than a thousand dollars in cash.' He wasn't screaming or anything like that, but he seemed disappointed in us. It was hard to hear that."

The lingering question for many SpaceX employees is when exactly they will see a big reward for all their work. SpaceX's staff is paid well but by no means exorbitantly. Many of them expect to make their money when SpaceX files for an initial public offering. The thing is that Musk does not want to go public anytime soon, and understandably so. It's a bit hard to explain the whole Mars thing to investors, when it's unclear what the business model around starting a colony on another planet will be. When the employees heard Musk say that an IPO was years away and would not occur until the Mars mission looked more secure, they started to grumble, and when Musk found out, he addressed all of SpaceX in an e-mail that is a fantastic window into his thinking and how it differs from almost every other CEO's. (The full e-mail appears in Appendix 3.)

June 7, 2013
Going Public

Per my recent comments, I am increasingly concerned about SpaceX going public before the Mars transport system is in place. Creating the technology needed to establish life on Mars is and always has been the fundamental goal of SpaceX. If being a public company diminishes that likelihood, then we should not do so until Mars is secure. This is something that I am open to reconsidering, but, given my experiences with Tesla and SolarCity, I am hesitant to foist being public on SpaceX, especially given the long term nature of our mission.

Some at SpaceX who have not been through a public company experience may think that being public is desirable. This is not so. Public company stocks, particularly if big step changes in technology are involved, go through extreme volatility, both for reasons of internal execution and for reasons that have nothing to do with anything except the economy. This causes people to be distracted by the manic-depressive nature of the stock instead of creating great products.

For those who are under the impression that they are so clever that they can outsmart public market investors and would sell SpaceX stock at the "right time," let me relieve you of any such notion. If you really are better than most hedge fund managers, then there is no need to worry about the value of your SpaceX stock, as you can just invest in other public company stocks and make billions of dollars in the market.

Elon

10

THE REVENGE OF THE ELECTRIC CAR

THERE ARE SO MANY TELEVISION COMMERCIALS FOR CARS AND TRUCKS that it's easy to become immune to them and ignore what's taking place in the ads. That's okay. Because there's not really much of note happening. Carmakers looking to put a modicum of effort into their ads have been hawking the exact same things for decades: a car with a bit more room, a few extra miles per gallon, better handling, or an extra cup holder. Those that can't find anything interesting at all to tout about their cars turn to scantily clad women, men with British accents, and, when necessary, dancing mice in tuxedos to try and convince people that their products are better than the rest. Next time a car ad appears on your television, pause for a moment and really listen to what's being said. When you realize that the Volkswagen sign-and-drive "event" is code for "we're making the experience of buying a car slightly less miserable than usual," you'll start to appreciate just how low the automotive industry has sunk.

In the middle of 2012, Tesla Motors stunned its complacent

peers in the automotive industry. It began shipping the Model S
sedan. This all-electric luxury vehicle could go more than 300
miles on a single charge. It could reach 60 miles per hour in
4.2 seconds. It could seat seven people, if you used a couple of
optional rear-facing seats in the back for kids. It also had two
trunks. There was the standard one and then what Tesla calls a
"frunk" up front, where the bulky engine would usually be. The
Model S ran on an electric battery pack that makes up the base
of the car and a watermelon-sized electric motor located between
the rear tires. Getting rid of the engine and its din of clanging
machinery also meant that the Model S ran silently. The Model S
outclassed most other luxury sedans in terms of raw speed, mileage, handling, and storage space.

And there was more—like a cutesy thing with the door handles, which were flush with the car's body until the driver got
close to the Model S. Then the silver handles would pop out, the
driver would open the door and get in, and the handles would
retract flush with the car's body again. Once inside, the driver
encountered a seventeen-inch touch-screen that controlled the
vast majority of the car's functions, be it raising the volume on
the stereo* or opening the sunroof with a slide of the finger.
Whereas most cars have a large dashboard to accommodate various displays and buttons and to protect people from the noise of
the engine, the Model S offered up vast amounts of space. The
Model S had an ever-present Internet connection, allowing the
driver to stream music through the touch console and to display
massive Google maps for navigation. The driver didn't need to
turn a key or even push an ignition button to start the car. His
weight in the seat coupled with a sensor in the key fob, which
is shaped like a tiny Model S, was enough to activate the vehi-

---

* The volume level on the sound system naturally goes to 11—an homage to *This Is Spinal Tap* and a reflection of Musk's sense of humor.

cle. Made of lightweight aluminum, the car achieved the highest safety rating in history. And it could be recharged *for free* at Tesla's stations lining highways across the United States and later around the world.

For both engineers and green-minded people, the Model S presented a model of efficiency. Traditional cars and hybrids have anywhere from hundreds to thousands of moving parts. The engine must perform constant, controlled explosions with pistons, crankshafts, oil filters, alternators, fans, distributors, valves, coils, and cylinders among the many pieces of machinery needed for the work. The oomph produced by the engine must then be passed through clutches, gears, and driveshafts to make the wheels turn, and then exhaust systems have to deal with the waste. Cars end up being about 10–20 percent efficient at turning the input of gasoline into the output of propulsion. Most of the energy (about 70 percent) is lost as heat in the engine, while the rest is lost through wind resistance, braking, and other mechanical functions. The Model S, by contrast, has about a dozen moving parts, with the battery pack sending energy instantly to a watermelon-sized motor that turns the wheels. The Model S ends up being about 60 percent efficient, losing most of the rest of its energy to heat. The sedan gets the equivalent of about 100 miles per gallon.*

Yet another distinguishing characteristic of the Model S was the experience of buying and owning the car. You didn't go to a dealership and haggle with a pushy salesman. Tesla sold the Model S directly through its own stores and website. Typically, the stores were placed in high-end malls or affluent suburbs, not

---

* And it's not just that the Model S and other electric cars are three to four times more efficient than internal combustion vehicles. They can also tap into power that is produced in centralized, efficient ways by power plants and solar arrays.

far from the Apple stores on which they were modeled. Customers would walk in and find a complete Model S in the middle of the shop and often an exposed version of the car's base near the back of the store to show off the battery pack and motor. There were massive touch-screens where people could calculate how much they might save on fuel costs by moving to an all-electric car, and where they could configure the look and add-ons for their future Model S. Once the configuration process was done, the customer could give the screen a big, forceful swipe and his Model S would theatrically appear on an even bigger screen in the center of the store. If you wanted to sit in the display model, a salesman would pull back a red velvet rope near the driver's-side door and let you enter the car. The salespeople were not compensated on commission and didn't have to try to talk you into buying a suite of extras. Whether you ultimately bought the car in the store or online, it was delivered in a concierge fashion. Tesla would bring it to your home, office, or anywhere else you wanted it. The company also offered customers the option of picking their cars up from the factory in Silicon Valley and treating their friends and family to a complimentary tour of the facility. In the months that followed the delivery, there were no oil changes or tune-ups to be dealt with because the Model S didn't need them. It had done away with so much of the mechanical dreck standard in an internal combustion vehicle. However, if something did go wrong with the car, Tesla would come pick it up and give the customer a loaner while it repaired the Model S.

The Model S also offered a way to fix issues in a manner that people had never before encountered with a mass-produced car. Some of the early owners complained about glitches like the door handles not popping out quite right or their windshield wipers operating at funky speeds. These were inexcusable flaws for such a costly vehicle, but Tesla typically moved with clever efficiency to address them. While the owner slept, Tesla's engineers tapped into the car via the Internet connection and downloaded software updates. When the customer took the car out for a spin in the morning and found it working right, he was left feeling as if magical elves had done the work. Tesla soon began showing off its software skills for jobs other than making up for mistakes. It put out a smartphone app that let people turn on their air-conditioning or heating from afar and to see where the car was parked on a map. Tesla also began installing software updates that imbued the Model S with new features. Overnight, the Model S sometimes got new traction controls for hilly and highway driving or could suddenly recharge much faster than before or possess a new range of voice controls. Tesla had transformed the car into a gadget—a device that actually got better after you bought it. As Craig Venter, one of the earliest Model S owners and the famed scientist who first decoded man's DNA, put it, "It changes everything about transportation. It's a computer on wheels."

The first people to notice what Tesla had accomplished were the technophiles in Silicon Valley. The region is filled with early adopters willing to buy the latest gizmos and suffer through their bugs. Normally this habit applies to computing devices ranging from $100 to $2,000 in price. This time around, the early adopters proved willing not only to spend $100,000 on a product that might not work but also to trust their well-being to a start-up. Tesla needed this early boost of confidence and got it on a scale few expected. In the first couple of months after the Model S went on sale, you might see one or two per day on the streets of San Francisco and the surrounding cities. Then you started to see five to ten per day. Soon enough, the Model S seemed to feel like the most common car in Palo Alto and Mountain View, the two cities at the heart of Silicon Valley. The Model S emerged as the ultimate status symbol for wealthy technophiles, allowing them to show off, get a new gadget, and claim to be helping the environment at the same time. From Silicon Valley, the Model S phenomenon spread to Los Angeles, then all along the West Coast and then to Washington, D.C., and New York (although to a lesser degree).

At first the more traditional automakers viewed the Model S as a gimmick and its surging sales as part of a fad. These sentiments, however, soon gave way to something more akin to panic. In November 2012, just a few months after it started shipping, the Model S was named *Motor Trend*'s Car of the Year in the first unanimous vote that anyone at the magazine could remember. The Model S beat out eleven other vehicles from companies such as Porsche, BMW, Lexus, and Subaru and was heralded as "proof positive that America can still make great things." *Motor Trend* celebrated the Model S as the first non–internal combustion engine car ever to win its top award and wrote that the vehicle handled like a sports car, drove as smoothly as a Rolls-Royce, held as much as a Chevy Equinox, and was more efficient than a Toyota Prius. Several months later, *Consumer Reports* gave the Model S its highest car rating in history—99 out of 100—while proclaiming that it was likely the best car ever built. It was at about this time that sales of the Model S started to soar alongside Tesla's share price and that General Motors, among other automakers, pulled together a team to study the Model S, Tesla, and the methods of Elon Musk.

It's worth pausing for a moment to meditate on what Tesla had accomplished. Musk had set out to make an electric car that did not suffer from any compromises. He did that. Then, using a form of entrepreneurial judo, he upended the decades of criticisms against electric cars. The Model S was not just the best electric car; it was best car, period, and *the* car people desired. America had not seen a successful car company since Chrysler emerged in 1925. Silicon Valley had done little of note in the automotive industry. Musk had never run a car factory before and was considered arrogant and amateurish by Detroit. Yet, one year after the Model S went on sale, Tesla had posted a profit, hit $562 million in quarterly revenue, raised its sales forecast, and become as valuable as Mazda Motor. Elon Musk had built the automotive equivalent of the iPhone. And car executives in Detroit, Japan, and Germany had only their crappy ads to watch as they pondered how such a thing had occurred.

You can forgive the automotive industry veterans for being caught unawares. For years Tesla had looked like an utter disaster incapable of doing much of anything right. It took until early 2009 for Tesla to really hit its stride with the Roadster and work out the manufacturing issues behind the sports car. Just as the company tried to build some momentum around the Roadster, Musk sent out an e-mail to customers declaring a price hike. Where the car originally started around $92,000, it would now start at $109,000. In the e-mail, Musk said that four hundred customers who had already placed their orders for a Roadster but not yet received them would bear the brunt of the price change and need to cough up the extra cash. He tried to assuage Tesla's customer base by arguing that the company had no choice but to raise prices. The manufacturing costs for the Roadster had come in much higher than the company initially expected, and Tesla needed to prove that it could make the cars at a profit to bolster its chances of securing a large government loan that would be needed to build the Model S, which it vowed to deliver in 2011. "I firmly believe that the plan . . . strikes a reasonable compromise between being fair to early customers and ensuring the viability of Tesla, which is obviously in the best interests of all customers," Musk wrote in the e-mail. "Mass market electric cars have been my goal from the beginning of Tesla. I don't want and I don't think the vast majority of Tesla customers want us to do anything to jeopardize that objective." While some Tesla customers grumbled, Musk had largely read his customer base right. They would support just about anything he suggested.

Following the price increase, Tesla had a safety recall. It said that Lotus, the manufacturer of the Roadster's chassis, had failed to tighten a bolt properly on its assembly line*. On the plus side, Tesla had only delivered about 345 Roadsters, which meant that it could fix the problem in a manageable fashion. On the downside, a safety recall was the last thing a car start-up needs, even if it was, as Tesla claimed, more of a proactive measure than anything else. The next year, Tesla had another voluntary recall. It had received a report of a power cable grinding against the body of the Roadster to the point that it caused a short circuit and some smoke. That time, Tesla brought 439 Roadsters in for a fix. Tesla did its best to put a positive spin on these issues, saying that it would make "house calls" to fix the Roadsters or pick up the cars and take them back to the factory. Ever since, Musk has tried to turn any snafu with a Tesla into an excuse to show off the company's attention to service and dedication to pleasing the customer. More often than not, the strategy has worked.

On top of the occasional issues with the Roadster, Tesla continued to suffer from public perception problems. In June 2009, Martin Eberhard sued Musk and went to town in the complaint detailing his ouster from the company. Eberhard accused Musk of libel, slander, and breach of contract. The charges painted Musk as a bully moneyman who had pushed the soulful inventor out of his own company. The lawsuit also accused Musk of trumping up his role in Tesla's founding. Musk responded in kind, issuing a blog post that detailed his take on Eberhard's foibles and taking

---

*Lotus had to recall some of its own cars for the same issue at the time.

umbrage at the suggestions that he was not a true founder of the company. A short while later, the two men settled and agreed to stop going at each other. "As co-founder of the company, Elon's contributions to Tesla have been extraordinary," Eberhard said in a statement at the time. It must have been excruciating for Eberhard to agree to put that in writing and the very existence of that statement points to Musk's skills and tactics as a hard-line negotiator. The two men continue to despise each other today, although they must do so in private, as legally required. Eberhard, though, holds no long-standing grudge against Tesla. His shares in the company ended up becoming very valuable. He still drives his Roadster, and his wife got a Model S.

For so much of its early existence, Tesla appeared in the news for the wrong reasons. There were people in the media and the automotive industry who viewed it as a gimmick. They seemed to delight in the soap opera–worthy spats between Musk and Eberhard and other disgruntled former employees. Far from being seen universally as a successful entrepreneur, Musk was viewed in some Silicon Valley circles as an abrasive blowhard who would get what he deserved when Tesla inevitably collapsed. The Roadster would make its way to the electric-car graveyard. Detroit would prove that it had a better handle on this whole car innovation thing than Silicon Valley. The natural order of the world would remain intact.

A funny thing happened, however. Tesla did just enough to survive. From 2008 to 2012, Tesla sold about 2,500 Roadsters.* The car had accomplished what Musk had intended from the out-

---

* When the very first Roadster arrived, it came in a large plywood crate. Tesla's engineers unpacked it furiously, installed the battery pack, and then let Musk take it for a spin. About twenty Tesla engineers jumped in prototype vehicles and formed a convoy that followed Musk around Palo Alto and Stanford.

set. It proved that electric cars could be fun to drive and that they could be objects of desire. With the Roadster, Tesla kept electric cars in the public's consciousness and did so under impossible circumstances, namely the collapse of the American automotive industry and the global financial markets. Whether Musk was a founder of Tesla in the purest sense of the word is irrelevant at this point. There would be no Tesla to talk about today were it not for Musk's money, marketing savvy, chicanery, engineering smarts, and indomitable spirit. Tesla was, in effect, willed into existence by Musk and reflects his personality as much as Intel, Microsoft, and Apple reflect the personalities of their founders. Marc Tarpenning, the other Tesla cofounder, said as much when he reflected on what Musk has meant to the company. "Elon pushed Tesla so much farther than we ever imagined," he said.

As difficult as birthing the Roadster had been, the adventure had whetted Musk's appetite for what he could accomplish in the automotive industry with a clean slate. Tesla's next car—code-named WhiteStar—would not be an adapted version of another company's vehicle. It would be made from scratch and structured to take full advantage of what the electric-car technology offered. The battery pack in the Roadster, for example, had to be placed near the rear of the car because of constraints imposed by the Lotus Elise chassis. This was okay but not ideal due to the imposing weight of the batteries. With WhiteStar, which would become the Model S, Musk and Tesla's engineers knew from the start that they would place the 1,300-pound battery pack on the base of the car. This would give the vehicle a low center of gravity and excellent handling. It would also give the Model S what's known as a low polar moment of inertia, which relates to how a car resists turning. Ideally, you want heavy parts like the engine as close as possible to the car's center of gravity, which is why the engines of race cars tend to be near the middle of the vehicle.

Traditional cars are a mess on this metric, with the bulky engine up front, passengers in the middle, and gasoline sloshing around the rear. In the case of the Model S, the bulk of the car's mass is very close to the center of gravity and this has positive follow-on effects to handling, performance, and safety.

The innards, though, were just one part of what would make the Model S shine. Musk wanted to make a statement with the car's look as well. It would be a sedan, yes, but it would be a sexy sedan. It would also be comfortable and luxurious and have none of the compromises that Tesla had been forced to embrace with the Roadster. To bring such a beautiful, functional car to life, Musk hired Henrik Fisker, a Danish automobile designer renowned for his work at Aston Martin.

Tesla first revealed its plans for the Model S to Fisker in 2007. It asked him to design a sleek, four-door sedan that would cost between $50,000 and $70,000. Tesla could still barely make Roadsters and had no idea if its all-electric powertrain would hold up over time. Musk, though, refused to wait and find out. He wanted the Model S to ship in late 2009 or early 2010 and needed Fisker to work fast. By reputation, Fisker had a flair for the dramatic and had produced some of the most stunning car designs over the past decade, not just for Aston Martin but also for special versions of BMW and Mercedes-Benz vehicles.

Fisker had a studio in Orange County, California, and Musk and other Tesla executives would meet there to go over his evolving takes on the Model S. Each visit was less inspiring than the last. Fisker baffled the Tesla teams with his stodgy designs. "Some of the early styles were like a giant egg," said Ron Lloyd, the former vice president of the WhiteStar project at Tesla. "They were terrible." When Musk pushed back, Fisker blamed the physical constraints Tesla had put in place for the Model S as too restrictive. "He said they would not let him make the car sexy," Lloyd said. Fisker tried a couple of different approaches and unveiled some foam models of the car for Musk and his crew to dissect. "We kept on telling him they were not right," Lloyd said.

Not long after these meetings, Fisker started his own company—Fisker Automotive—and unveiled the Fisker Karma hybrid in 2008. This luxury sedan looked like a vehicle Batman might take out for a Sunday drive. With its elongated lines and sharp edges, the car was stunning and truly original. "It rapidly became clear that he was trying to compete with us," Lloyd said. As Musk dug into the situation, he discovered that Fisker had been shopping his idea for a car company to investors around Silicon Valley for some time. Kleiner Perkins Caufield & Byers, one of the more famous venture capital firms in Silicon Valley, once had a chance to invest in Tesla and then ended up putting money into Fisker instead. All of this was too much for Musk, and he launched a lawsuit against Fisker in 2008, accusing him of stealing Tesla's ideas and using the $875,000 Tesla had paid for design work to help get his rival car company off the ground. Fisker ultimately prevailed in the dispute with an arbitrator ordering Tesla to reimburse Fisker's legal fees and deeming Tesla's allegations baseless.

Tesla had thought about doing a hybrid like Fisker where a gas engine would be present to recharge the car's batteries after they had consumed an initial charge. The car would be able to travel fifty to eighty miles after being plugged into an outlet and then take advantage of ubiquitous gas stations as needed to top up the batteries, eliminating range anxiety. Tesla's engineers prototyped the hybrid vehicle and ran all sorts of cost and performance metrics. In the end, they found the hybrid to be too much of a compromise. "It would be expensive, and the performance would not be as good as the all-electric car," said J. B. Straubel. "And we would have needed to build a team to compete with the core competency of every car company in the world. We would have been betting against all the things we believe in, like the power electronics and batteries improving. We decided to put all the effort into going where we think the endpoint is and to never look back." After coming to this conclusion, Straubel and others inside Tesla started to let go of their anger toward Fisker. They figured he would end up delivering a kluge of a car and get what was coming to him.

A large car company might spend $1 billion and need thousands of people to design a new vehicle and bring it to market. Tesla had nothing close to these resources as it gave birth to the Model S. According to Lloyd, Tesla initially aimed to make about ten thousand Model S sedans per year and had budgeted around $130 million to achieve this goal, including engineering the car and acquiring the manufacturing machines needed to stamp out the body parts. "One of the things Elon pushed hard with everyone was to do as much as possible in-house," Lloyd said. Tesla would make up for its lack of R&D money by hiring smart people who could outwork and outthink the third parties relied on by the rest of the automakers. "The mantra was that one great engineer will replace three medium ones," Lloyd said.

A small team of Tesla engineers began the process of trying to figure out the mechanical inner workings of the Model S. Their first step in this journey took place at a Mercedes dealership where they test drove a CLS 4-Door Coupe and an E-Class sedan. The cars had the same chassis, and the Tesla engineers took measurements of every inch of the vehicles, studying what they liked and didn't like. In the end, they preferred the styling on the CLS and settled on it as their baseline for thinking about the Model S.

After purchasing a CLS, Tesla's engineers tore it apart. One team had reshaped the boxy, rectangular battery pack from the

Roadster and made it flat. The engineers cut the floor out of the CLS and plopped in the pack. Next they put the electronics that tied the whole system together in the trunk. After that, they replaced the interior of the car to restore its fit and finish. Following three months of work, Tesla had in effect built an all-electric Mercedes CLS. Tesla used the car to woo investors and future partners like Daimler that would eventually turn to Tesla for electric powertrains in their vehicles. Now and again, the Tesla team took the car out for drives on public roads. It weighed more than the Roadster but was still fast and had a range of about 120 miles per charge. To perform these joyrides-cum-tests in relative secrecy, the engineers had to weld the tips of the exhaust pipes back onto the car to make it look like any other CLS.

It was at this time, the summer of 2008, when an artsy car lover named Franz von Holzhausen joined Tesla. His job would be to breathe new life into the car's early designs and, if possible, turn the Model S into an iconic product.*

Von Holzhausen grew up in a small Connecticut town. His father worked on the design and marketing of consumer products, and Franz treated the family basement full of markers, different kinds of paper, and other materials as a playground for his imagination. As he grew older, von Holzhausen drifted toward cars.

---

* At some point from late 2007 to 2008, Musk also tried to hire Tony Fadell, an executive at Apple who is credited with bringing the iPod and iPhone to life. Fadell remembered being recruited for the CEO job at Tesla, while Musk remembered it more as a chief operating officer type of position. "Elon and I had multiple discussions about me joining as Tesla's CEO, and he even went to the lengths of staging a surprise party for me when I was going to visit their offices," Fadell said. Steve Jobs caught wind of these meetings and turned on the charm to keep Fadell. "He was sure nice to me for a while," Fadell said. A couple of years later, Fadell left Apple to found Nest, a maker of smart-home devices, which Google then acquired in 2014.

He and a friend stripped down a dune-buggy motor one winter and then built it back up, and von Holzhausen always filled the margins of his school notebooks with drawings of cars and had pictures of cars on his bedroom walls. Applying to college, von Holzhausen decided to follow his father's path and enrolled in the industrial design program at Syracuse University. Then, through a chance encounter with another designer during an internship, von Holzhausen heard about the Art Center College of Design in Los Angeles. "This guy had been teaching me about car design and this school in Los Angeles, and I got super-intrigued," said von Holzhausen. "I went to Syracuse for two years and then decided to transfer out to California."

The move to Los Angeles kicked off a long and storied design career in the automotive industry. Von Holzhausen would go on to intern in Michigan with Ford and in Europe with Volkswagen, where he began to pick up on a mix of design sensibilities. After graduating in 1992, he started work for Volkswagen on just about the most exciting project imaginable—a top-secret new version of the Beetle. "It really was a magical time," von Holzhausen said. "Only fifty people in the world knew we were doing this project." Von Holzhausen had a chance to work on the exterior and interior of the vehicle, including the signature flower vase built into the dashboard. In 1997, Volkswagen launched the "New Beetle," and von Holzhausen saw firsthand how the look of the car captivated the public and changed the way people felt about Volkswagen, which had suffered from woeful sales in the United States. "It started a rebirth of the VW brand and brought design back into their mix," he said.

Von Holzhausen spent eight years with VW, climbing the ranks of its design team and falling in love with the car culture of Southern California. Los Angeles has long adored its cars, with the climate lending itself to all manner of vehicles from convert-

ibles to surfboard-toting vans. Almost all of the major carmakers set up design studios in the city. The presence of the studios
allowed von Holzhausen to hop from VW to General Motors and
Mazda, where he served as the company's director of design.

GM taught von Holzhausen just how nasty a big car company
could become. None of the cars in GM's lineup really excited
him, and it seemed near impossible to make a large impact on the
company's culture. He was one member of a thousand-person
design team that divvyed up the makes of cars haphazardly
without any consideration as to which person *really* wanted to
work on which car. "They took all the spirit out of me," said
von Holzhausen. "I knew I didn't want to die there." Mazda, by
contrast, needed and wanted help. It let von Holzhausen and his
team in Los Angeles put their imprint on every car in the North
American vehicle lineup and to produce a set of concept cars that
reshaped how the company approached design. As von Holzhausen put it, "We brought the zoom-zoom back into the look and
feel of the car."

Von Holzhausen started a project to make Mazda's cars more
green by revaluating the types of materials used to fabricate the
seats and the fuels going into the vehicles. He had, in fact, just
made an ethanol-based concept car when, in early 2008, a friend
told him that Tesla needed a chief designer. After playing phone
tag for a month with Musk's assistant, Mary Beth Brown, to
inquire about the position, von Holzhausen finally got in touch
and met Musk for an interview at the SpaceX headquarters.

Musk instantly saw von Holzhausen, with his bouffant,
trendy clothes and laid-back attitude, as a free-spirited, creative
complement and wooed him with vigor. They took a tour of the
SpaceX factory in Hawthorne and Tesla's headquarters in Silicon Valley. Both facilities were chaotic and reeked of start-up.
Musk ramped up the charm and sold von Holzhausen on the

idea that he had a chance to shape the future of the automobile and that it made sense to leave his cushy job at a big, proven automaker for this once-in-a-lifetime opportunity. "Elon and I went for a drive in the Roadster, and everyone was checking it out," von Holzhausen said. "I knew I could stay at Mazda for ten years and get very comfortable or take a huge leap of faith. At Tesla, there was no history, no baggage. There was just a vision of products that could change the world. Who wouldn't want to be involved with that?"

While von Holzhausen knew the risks of going to a start-up, he could not have realized just how close Tesla was to bankruptcy when he joined the company in August 2008. Musk had coaxed von Holzhausen away from a secure job and into the jaws of death. But in many ways, this is what von Holzhausen sought at this point in his career. Tesla did not feel as much like a car company as a bunch of guys tinkering on a big idea. "To me, it was exciting," he said. "It was like a garage experiment, and it made cars cool again." The suits were gone, and so were the veteran automotive hands dulled by years working in the industry. In their stead, von Holzhausen found energetic geeks who didn't realize that what they wanted to do was borderline impossible. Musk's presence added to the energy and gave von Holzhausen confidence that Tesla actually could outflank much, much larger competitors. "Elon's mind was always way beyond the present moment," he said. "You could see that he was a step or three ahead of everyone else and one hundred percent committed to what we were doing."

Von Holzhausen had examined the drawings of the Model S left by Fisker and a clay model of the car and had come away unimpressed. "It was a blob," he said. "It was clear to me that the people that had been working on this were novices." Musk realized the same thing and tried to articulate what he wanted. Even though the words were not precise, they were good enough to give von Holzhausen a feel for Musk's vision and the confidence that he could deliver on it. "I said, 'We're going to start over. We're going to work together and make this awesome.'"

To save money, the Tesla design center came to life inside the SpaceX factory. A handful of people on von Holzhausen's team took over one corner and put up a tent to add some separation and secrecy to what they were doing. In the tradition of many a Musk employee, von Holzhausen had to build his own office. He made a pilgrimage to IKEA to buy some desks and then went to an art store to get some paper and pens.

As von Holzhausen began sketching the outside of the Model S, the Tesla engineers had started up a project to build another electric CLS. They ripped this one down to its very core, removing all of the body structure and then stretching the wheelbase by four inches to match up with some of the early Model S specifications. Things began moving fast for everyone involved in the Model S project. In the span of about three months, von Holzhausen had designed 95 percent of what people see today with the Model S, and the engineers had started building a prototype exterior around the skeleton.

Throughout this process, von Holzhausen and Musk talked every day. Their desks were close, and the men had a natural rapport. Musk said he wanted an aesthetic that borrowed from Aston Martin and Porsche and some specific functions. He insisted, for example, that the car seat seven people. "It was like 'Holy shit, how do we pull this off in a sedan?'" von Holzhausen said. "But I understood. He had five kids and wanted something that could be thought of as a family vehicle, and he knew other people would have this issue."

Musk wanted to make another statement with a huge touchscreen. This was years before the iPad would be released. The touch-screens that people ran into now and again at airports or shopping kiosks were for the most part terrible. But to Musk, the iPhone and all of its touch functions made it obvious that this type of technology would soon become commonplace. He would make a giant iPhone and have it handle most of the car's functions. To find the right size for the screen, Musk and von Holzhausen would sit in the skeleton car and hold up laptops of different sizes, placing them horizontally and vertically to see what looked best. They settled on a seventeen-inch screen in a vertical position. Drivers would tap on this screen for every task except for opening the glove box and turning on the emergency lights—jobs required by law to be performed with physical buttons.

Since the battery pack at the base of the car would weigh so much, Musk, the designers, and the engineers were always looking for ways to reduce the Model S's weight in other spots. Musk opted to solve a big chunk of this problem by making the body of the Model S out of lightweight aluminum instead of steel. "The non-battery-pack portion of the car has to be lighter than comparable gasoline cars, and making it all aluminum became the obvious decision," Musk said. "The fundamental problem was that if we didn't make it out of aluminum the car wasn't going to be any good."

Musk's word choice there—"obvious decision"—goes a long way toward explaining how he operates. Yes, the car needed to be light, and, yes, aluminum would be an option for making that happen. But at the time, car manufacturers in North America had almost no experience producing aluminum body panels. Aluminum tends to tear when worked by large presses. It also develops lines that look like stretch marks on skin and make it difficult to lay down smooth coats of paint. "In Europe, you had some Jaguars and one Audi that were made of aluminum, but it was less than five percent of the market," Musk said. "In North America, there was nothing. It's only recently that the Ford F-150 has arrived as mostly aluminum. Before that, we were the only one." Inside of Tesla, attempts were repeatedly made to talk Musk out of the aluminum body, but he would not budge, seeing it as the only rational choice. It would be up to the Tesla team to figure out how to make the aluminum manufacturing happen. "We knew it could be done," Musk said. "It was a question of how hard it would be and how long it would take us to sort it out."

Just about all of the major design choices with the Model S came with similar challenges. "When we first talked about the touch-screen, the guys came back and said, 'There's nothing like that in the automotive supply chain,'" Musk said. "I said, 'I know. That's because it's never been put in a fucking car before.'" Musk figured that computer manufacturers had tons of experience making seventeen-inch laptop screens and expected them to knock out a screen for the Model S with relative ease. "The laptops are pretty robust," Musk said. "You can drop them and leave them out in the sun, and they still have to work." After contacting the laptop suppliers, Tesla's engineers came back and said that the temperature and vibration loads for the computers did not appear to be up to automotive standards. Tesla's supplier in Asia also kept pointing the carmaker to its automotive division instead of its computing division. As Musk dug into the situation more, he discovered that the laptop screens simply had not been tested before under the tougher automotive conditions, which included large temperature fluctuations. When Tesla performed the tests, the electronics ended up working just fine. Tesla also started working hand in hand with the Asian manufacturers to perfect their then-immature capacitive-touch technology and to find ways to hide the wiring behind the screen that made the touch technology possible. "I'm pretty sure that we ended up with the only seventeen-inch touch-screen in the world," Musk said. "None of the computer makers or Apple had made it work yet."

The Tesla engineers were radical by automotive industry standards but even they had problems fully committing to Musk's vision. "They wanted to put in a bloody switch or a button for the lights," Musk said. "Why would we need a switch? When it's dark, turn the lights on." Next, the engineers put up resistance to the door handles. Musk and von Holzhausen had been studying a bunch of preliminary designs in which the handles had yet to be drawn in and started to fall in love with how clean the car looked. They decided that the handles should only present themselves when a passenger needed to get in the car. Right away, the engineers realized this would be a technological pain, and they completely ignored the idea in one prototype version of the car, much to the dismay of Musk and von Holzhausen. "This prototype had the handles pivot instead of popping out," von Holzhausen said. "I was upset about it, and Elon said, 'Why the fuck is this different? We're not doing this.'"

To crank up the pace of the Model S design, there were engineers working all day and then others who would show up at 9 P.M. and work through the night. Both groups huddled inside of the 3,000-square-foot tent placed on the SpaceX factory floor. Their workspace looked like a reception area at an outdoor wedding. "The SpaceX guys were amazingly respectful and didn't peek or ask questions," said Ali Javidan, one of the main engineers. As von Holzhausen delivered his specifications, the engineers built the prototype body of the car. Every Friday afternoon, they brought what they had made into a courtyard behind the factory where Musk would look it over and provide feedback. To run tests on the body, the car would be loaded up with ballast to represent five people and then do loops around the factory until it overheated or broke down.

The more von Holzhausen learned about Tesla's financial struggles, the more he wanted the public to see the Model S.

"Things were so precarious, and I didn't want to miss our opportunity to get this thing finished and show it to the world," he said. That moment came in March 2009, when, just six months after von Holzhausen had arrived, Tesla unveiled the Model S at a press event held at SpaceX.

Amid rocket engines and hunks of aluminum, Tesla showcased a gray Model S sedan. From a distance, the display model looked glamorous and refined. The media reports from the day described the car as the love child of an Aston Martin and a Maserati. In reality, the sedan barely held together. It still had the base structure of a Mercedes CLS, although no one in the press knew that, and some of the body panels and the hood were stuck to the frame with magnets. "They could just slide the hood right off," said Bruce Leak, a Tesla owner invited to attend the event. "It wasn't really attached. They would put it back on and try and align it to get the fit and finish right, but then someone would push on it, and it would move again. It was one of those Wizard of Oz, man behind the curtain moments." A couple of the Tesla engineers practiced test-driving the car for a couple of days leading up to the event to make sure that they knew just how long the car would go before it overheated. While not perfect, the display accomplished exactly what Musk had intended. It reminded people that Tesla had a credible plan to make electric cars more mainstream and that its cars were far more ambitious than what big-time automakers like GM and Nissan seemed to have in mind both from a design and a range perspective.

The messy reality behind the display was that the odds of Tesla advancing the Model S from a prop to a sellable car were infinitesimal. The company had the technical know-how and the will for the job. It just didn't have much money or a factory that could crank out cars by the thousands. Building an entire car would require blanking machines that take sheets of aluminum and chop them up into the appropriate size for doors, hoods, and body panels. Next up would be the massive stamping machines and metal dies used to take the aluminum and bend it into precise shapes. Then there would be dozens of robots that would aid in assembling the cars, computer-controlled milling machines for precise metalwork, painting equipment, and a bevy of other machines for running tests. It was an investment that would run into the hundreds of millions of dollars. Musk would also need to hire thousands of workers.

As with SpaceX, Musk preferred to build as much of Tesla's vehicles in-house as possible, but the high costs were limiting just how much Tesla could take on. "The original plan was that we would do final assembly," said Diarmuid O'Connell, the vice president of business development at Tesla. Partners would stamp out the body parts, do the welding and handle the painting, and ship everything to Tesla, where workers would turn the parts into a whole car. Tesla proposed to build a factory to handle this type of work first in Albuquerque, New Mexico, and then later in San Jose, California, and then pulled back on these proposals, much to the dismay of city officials in both locales. The public hemming and hawing around picking the factory site did little to inspire confidence in Tesla's ability to knock out a second car and generated the same type of negative headlines that had surrounded the Roadster's protracted delivery.

O'Connell had joined Tesla in 2006 to help solve some of the factory and financing issues. He grew up near Boston in a middle-class Irish family and went on to earn a bachelor's degree from Dartmouth College. After that, O'Connell attended the University of Virginia to get a master's degree in foreign policy and then Northwestern, where he got an MBA from the Kellogg School of Management. He had fancied himself a scholar of the Soviet Union and its foreign and economic policy and had studied these areas at UVa. "But then, in 1988 and 1989, they're starting to close down the Soviet Union, and, at the very least, I had a brand problem," O'Connell said. "It started looking to me like I was heading to a career in academia or intelligence." It was then that O'Connell's career took a detour into the business world, where he became a management consultant working for McCann Erickson Worldwide, Young & Rubicam, and Accenture, advising companies like Coca-Cola and AT&T.

O'Connell's career path changed more drastically in 2001 when the planes hit the twin towers in New York. In the wake of the terrorist attacks, O'Connell, like many people, decided to serve the United States in any capacity that he could. In his late thirties, he had missed the window to be a soldier and instead focused his attention on trying to get into national security work. O'Connell went from office to office in Washington, D.C., looking for a job and had little luck until Lincoln Bloomfield, the assistant secretary of state for political-military affairs, heard him out. Bloomfield needed someone who could help prioritize missions in the Middle East and make sure the right people were working on the right things, and he figured that O'Connell's management consulting experience made him a nice fit for the job. O'Connell became Bloomfield's chief of staff and dealt with a wide range of charged situations, from trade negotiations to setting up an embassy in Baghdad. After gaining security clearance, O'Connell also had access to a daily report that collected information from intelligence and military personnel on the status of operations in Iraq and Afghanistan. "Every morning at six A.M., the first thing to hit my desk was this overnight report that included information on who got killed and what killed them," O'Connell said. "I kept thinking, This is insane. Why are we in this place? It was not just Iraq but the whole picture. Why were we so invested in that part of the world?" The unsurprising answer that O'Connell came up with was oil.

The more O'Connell dug into the United States' dependence on foreign oil, the more frustrated and despondent he became. "My clients were basically the combat commanders—people in charge of Latin America and Central Command," he said. "As I talked with them and studied and researched, I realized that even in peacetime, so many of our assets were employed to support the economic pipeline around oil." O'Connell decided that the rational thing to do for his country and for his newborn son was to alter this equation. He looked at the wind industry and the solar industry and the traditional automakers but came away unconvinced that what they were doing could have a radical enough impact on the status quo. Then, while reading *Businessweek*, he stumbled on an article about a start-up called Tesla Motors and went to the company's website, which described Tesla as a place "where we are doing things, not talking about things." "I sent an e-mail telling them I had come from the national security area and was really passionate about reducing our dependence on oil and figured it was just a dead-letter type of thing," O'Connell said. "I got an e-mail back the next day."

Musk hired O'Connell and quickly dispatched him to Washington, D.C., to start poking around on what types of tax credits and rebates Tesla might be able to drum up around its electric vehicles. At the same time, O'Connell drafted an application for a Department of Energy stimulus package.* "All I knew is that we were going to need a shitload of money to build this company," O'Connell said. "My view was that we needed to explore

---

* It took a couple of years, from about 2007 to 2009, for the Energy Department application to morph into the actual possibility of a loan from the government.

everything." Tesla had been looking for between $100 million and $200 million, grossly underestimating what it would take to build the Model S. "We were naïve and learning our way in the business," O'Connell said.

It January 2009, Tesla took over Porsche's usual spot at the Detroit auto show, getting the space cheap because so many other car companies had bailed out on the event. Fisker had a luxurious booth across the hallway with wood flooring and pretty blond booth babes draped over its car. Tesla had the Roadster, its electric powertrain, and no frills.

The technology that Tesla's engineers displayed proved good enough to attract the attention of the big boys. Not long after the show, Daimler voiced some interest in seeing what an electric Mercedes A Class car might look and feel like. Daimler executives said they would visit Tesla in about a month to discuss this proposition in detail, and the Tesla engineers decided to blow them away by producing two prototype vehicles before the visit. When the Daimler executives saw what Tesla had done, they ordered four thousand of Tesla's battery packs for a fleet of test vehicles in Germany. The Tesla team pulled off the same kind of feats for Toyota and won its business, too.

In May 2009, things started to take off for Tesla. The Model S had been unveiled, and Daimler followed that by acquiring a 10 percent stake in Tesla for $50 million. The companies also formed a strategic partnership to have Tesla provide the battery packs for one thousand of Daimler's Smart cars. "That money was important and went a long way back then," said O'Connell. "It was also a validation. Here is the company that invented the internal combustion engine, and they are investing in us. It was a seminal moment, and I am sure it gave the guys over at the DOE the feeling that we were real. It's not just our scientists saying this stuff is good. It's Mercedes freaking Benz."

Sure enough, in January 2010, the Department of Energy struck a $465 million loan agreement with Tesla.* The money was far more than Tesla had ever expected to get from the government. But it still represented just a fraction of the $1 billion plus that most carmakers needed to bring a new vehicle to market. So, while Musk and O'Connell were thrilled to get the money, they still wondered if Tesla would be able to live up to the bargain. Tesla would need one more windfall or, perhaps, to steal a car factory. And in May 2010, that's more or less what it did.

General Motors and Toyota had teamed up in 1984 to build New United Motor Manufacturing Inc., or NUMMI, on the site of a former GM assembly plant in Fremont, California, a city on the outskirts of Silicon Valley. The companies hoped the joint facility would combine the best of American and Japanese auto-making skills and result in higher-quality, cheaper cars. The factory went on to pump out millions of vehicles like the Chevy Nova and Toyota Corolla. Then the recession hit, and GM found itself trying to climb out of bankruptcy. It decided to abandon the plant in 2009, and Toyota followed right after, saying it would close down the whole facility, leaving five thousand people without jobs.

All of a sudden, Tesla had the chance to buy a 5.3-million-square-foot plant in its backyard. Just one month after the last Toyota Corolla went off the manufacturing line in April 2010, Tesla and Toyota announced a partnership and transfer of the factory. Tesla agreed to pay $42 million for a large portion of the factory (once worth $1 billion), while Toyota invested $50 million in Tesla for a 2.5 percent stake in the company. Tesla had

---

*   The deal had two parts. Tesla would keep making battery packs and associated technology that other companies might use, and it would produce its own electric vehicles at a manufacturing facility in the United States.

basically secured a factory, including the massive metal-stamping machines and other equipment, for free.*

The string of fortunate turns for Tesla left Musk feeling good. Just after the factory deal closed in the summer of 2010, Tesla started the process of filing for an initial public offering. The company obviously needed as much capital as it could get to bring the Model S to market and push forward with its other technology projects. Tesla hoped to raise about $200 million.

For Musk, going public represented something of a Faustian bargain. Ever since the Zip2 and PayPal days, Musk has done everything in his power to maintain absolute control over his companies. Even if he remained the largest shareholder in Tesla, the company would be subjected to the capricious nature of the public markets. Musk, the ultimate long-term thinker, would face constant second-guessing from investors looking for short-term returns. Tesla would also be subject to public scrutiny, as it would be forced to open its books for public consumption. This was bad because Musk prefers to operate in secrecy and because Tesla's financial situation looked awful. The company had one product (the Roadster), had huge development costs, and had bordered on bankruptcy months earlier. The car blog Jalopnik greeted the Tesla IPO as a Hail Mary rather than a sound fiscal move. "For lack of a better phrase, Tesla is a money pit," the blog wrote.

---

* Musk had received a lot of pushback internally for trying to locate a car factory in or near California. "All the guys in Detroit said it needs to be in a place where the labor can afford to live and be happy," Lloyd said. "There's a lot of learned skill on an assembly line, and you can't afford turnover." Musk responded that SpaceX had found a way to build rockets in Los Angeles, and that Tesla would find a way to build cars in Northern California. His stubbornness ended up being fortuitous for the company. "If it hadn't been for that DOE loan, and the NUMMI plant, there's no way Tesla would have ended up being so successful, so fast," Lloyd said.

"Since the company's founding in 2003, it's managed to incur over $290 million in losses on just $147.6 million in revenue." Told by a source that Tesla hoped to sell 20,000 units of the Model S per year at $58,000 a pop, Jalopnik scoffed. "Even considering the supposed pent-up demand among environmentalists for a car like the Model S, those are ambitious goals for a small company planning to launch a niche luxury product into a soft market. Frankly, we're skeptical. We've seen how brutal and unforgiving the market can be, and other automakers aren't simply going to roll over and surrender that volume to Tesla." Other pundits concurred with this assessment.

Tesla went public on June 29, 2010, nonetheless. It raised $226 million, with the company's shares shooting up 41 percent that day. Investors looked past Tesla's $55.7 million loss in 2009 and the more than $300 million the company had spent in seven years. The IPO stood as the first for an American carmaker since Ford went public in 1956. Competitors continued to treat Tesla like an annoying, ankle-biting dachshund. Nissan's CEO, Carlos Ghosn, used the event to remind people that Tesla was but a pipsqueak and that his company had plans to pump out up to 500,000 electric cars by 2012.

Flush with funds, Musk began expanding some of the engineering teams and formalizing the development work around the Model S. Tesla's main offices moved from San Mateo to a larger building in Palo Alto, and von Holzhausen expanded the design team in Los Angeles. Javidan hopped between projects, helping develop technology for the electrified Mercedes-Benz, an electric Toyota Rav4, and prototypes of the Model S. The Tesla team worked fast inside of a tiny lab with about 45 people knocking out 35 Rav4 test vehicles at the rate of about two cars per week. The alpha version of the Model S, including newly stamped body parts from the Fremont factory, a revamped battery pack, and revamped power electronics, came to life in the basement of the Palo Alto office. "The first prototype was finished at about two A.M.," Javidan said. "We were so excited that we drove it around without glass, any interior, or a hood."

A day or two later, Musk came to check out the vehicle. He jumped into the car and drove it to the opposite end of the basement, where he could spend some time alone with it. He got out and walked around the vehicle, and then the engineers came over to hear his take on the machine. This process would be repeated many times in the months to come. "He would generally be positive but constructive," Javidan said. "We would try and get him rides whenever we could, and he might ask for the steering to be tighter or something like that before running off to another meeting."

About a dozen of the alpha cars were produced. A couple went to suppliers like Bosch to begin work on the braking systems, while others were used for various tests and design tweaks. Tesla's executives kept the vehicles rotating on a strict schedule, giving one team two weeks for cold-weather testing and then shipping that alpha car to another team right away for powertrain tuning. "The guys from Toyota and Daimler were blown away," Javidan said. "They might have two hundred alpha cars and several hundred to a thousand beta cars. We were doing everything from crash tests to the interior design with about fifteen cars. That was amazing to them."

Tesla employees developed similar techniques to their counterparts at SpaceX for dealing with Musk's high demands. The savvy engineers knew better than to go into a meeting and deliver bad news without some sort of alternative plan at the ready. "One of the scariest meetings was when we needed to ask Elon for an extra two weeks and more money to build out another version of the Model S," Javidan said. "We put together a plan, stating how long things would take and what they would cost. We told him that if he wanted the car in thirty days it would require hiring some new people, and we presented him with a stack of resumes. You don't tell Elon you can't do something. That will get you kicked out of the room. You need everything lined up. After we presented the plan, he said, 'Okay, thanks.' Everyone was like, 'Holy shit, he didn't fire you.'"

There were times when Musk would overwhelm the Tesla engineers with his requests. He took a Model S prototype home for a weekend and came back on the Monday asking for around eighty changes. Since Musk never writes anything down, he held all the alterations in his head and would run down the checklist week by week to see what the engineers had fixed. The same engineering rules as those at SpaceX applied. You did what Musk asked or were prepared to burrow down into the properties of materials to explain why something could not be done. "He always said, 'Take it down to the physics,'" Javidan said.

As the development of the Model S neared completion in 2012, Musk refined his requests and dissection style. He went over the Model S with von Holzhausen every Friday at Tesla's design studio in Los Angeles. Von Holzhausen and his small team had moved out of the corner in the SpaceX factory and gotten their own hangar-shaped facility near the rear of the SpaceX complex.* The building had a few offices and then one large, wide-open area where various mock-ups of vehicles and parts awaited inspection. During a visit I made in 2012, there was one complete Model S, a skeletal version of the Model X—an as yet to be released SUV—and a selection of tires and hubcaps lined up against the wall. Musk sank into the Model S driver seat and von Holzhausen climbed into the passenger seat. Musk's eyes darted

---

* Boeing used to make fuselages for the 747 in the SpaceX building and painted them in what became the Tesla design studio.

around for a few moments and then settled onto the sun visor. It was beige and a visible seam ran around the edge and pushed the fabric out. "It's fish-lipped," Musk said. The screws attaching the visor to the car were visible as well, and Musk insisted that every time he saw them it felt like tiny daggers were stabbing him in the eyes. The whole situation was unacceptable. "We have to decide what is the best sun visor in the world and then do better," Musk said. A couple of assistants taking notes outside of the car jotted this down.

This process played out again with the Model X. This was to be Tesla's merger of an SUV and a minivan built off the Model S foundation. Von Holzhausen had four different versions of the vehicle's center console resting on the floor, so that they could be slotted in one by one and viewed by Musk. The pair spent most of their time, however, agonizing over the middle row of seats. Each one had an independent base so that each passenger could adjust his seat rather than moving the whole row collectively. Musk loved the freedom this gave the passenger but grew concerned after seeing all three seats in different positions. "The problem is that they will never be aligned and might look a mess," Musk said. "We have to make sure they are not too hodgy podgy."

The idea of Musk as a design expert has long struck me as bizarre. He's a physicist at heart and an engineer by demeanor. So much of who Musk is says that he should fall into that Silicon Valley stereotype of the schlubby nerd who would only know good design if he read about it in a textbook. The truth is that there might be some of that going on with Musk, and he's turned it into an advantage. He's very visual and can store things that others have deemed to look good away in his brain for recall at any time. This process has helped Musk develop a good eye, which he's combined with his own sensibilities, while also refining his ability to put what he wants into words. The result is a confident, assertive perspective that does resonate with the tastes of consumers. Like Steve Jobs before him, Musk is able to think up things that consumers did not even know they wanted—the door handles, the giant touch-screen—and to envision a shared point of view for all of Tesla's products and services. "Elon holds Tesla up as a product company," von Holzhausen said. "He's passionate that you have to get the product right. I have to deliver for him and make sure it's beautiful and attractive."

With the Model X, Musk again turned to his role as a dad to shape some of the flashiest design elements of the vehicle. He and von Holzhausen were walking around the floor of an auto show in Los Angeles, and they both complained about the awkwardness of getting to the middle and back row seats in an SUV. Parents who have felt their backs wrench while trying to angle a child and car seat into a vehicle know this reality all too well, as does any decent-sized human who has tried to wedge into a third row seat. "Even on a minivan, which is supposed to have more room, almost one-third of the entry space is covered by the sliding door," von Holzhausen said. "If you could open up the car in a way that is unique and special, that could be a real game changer. We took that kernel of an idea back and worked up forty or fifty design concepts to solve the problem, and I think we ended up with one of the most radical ones." The Model X has what Musk coined as "falcon-wing doors." They're hinged versions of the gull-wing doors found on some high-end cars like the DeLorean. The doors go up and then flop over in a constrained enough way that the Model X won't rub up against a car parked close to it or hit the ceiling in a garage. The end result is that a parent can plop a child in the second-row passenger seat without needing to bend over or twist at all.

When Tesla's engineers first heard about the falcon-wing doors, they cringed. Here was Musk with another crazy ask.

"Everyone tried to come up with an excuse as to why we couldn't do it," Javidan said. "You can't put it in the garage. It won't work with things like skis. Then, Elon took a demo model to his house and showed us that the doors opened. Everyone is mumbling, 'Yeah, in a fifteen-million-dollar house, the doors will open just fine.'" Like the controversial door handles on the Model S, the Model X's doors have become one of its most striking features and the thing consumers talk about the most. "I was one of the first people to test it out with a kid's car seat," Javidan said. "We have a minivan, and you have to be a contortionist to get the seat into the middle row. Compared to that, the Model X was so easy. If it's a gimmick, it's a gimmick that works."

During my 2012 visit to the design studio, Tesla had a number of competitors' vehicles in the parking lot nearby, and Musk made sure to demonstrate the limitations of their seating compared to the Model X. He tried with honest effort to sit in the third row of an Acura SUV, but, even though the car claimed to have room for seven, Musk's knees were pressed up to his chin, and he never really fit into the seat. "That's like a midget cave," he said. "Anyone can make a car big on the outside. The trick is to make it big on the inside." Musk went from one rival's car to the next, illuminating the vehicles' flaws for me and von Holzhausen. "It's good to get a sense for just how bad the other cars are," he said.

When these statements fly out of Musk's mouth, it's momentarily shocking. Here's a guy who needed nine years to produce about three thousand cars ridiculing automakers that build millions of vehicles every year. In that context, his ribbing comes off as absurd.

Musk, though, approaches everything from a Platonic perspective. As he sees it, all of the design and technology choices should be directed toward the goal of making a car as close to perfect as possible. To the extent that rival automakers haven't, that's what Musk is judging. It's almost a binary experience for him. Either you're trying to make something spectacular with no compromises or you're not. And if you're not, Musk considers you a failure. This position can look unreasonable or foolish to outsiders, but the philosophy works for Musk and constantly pushes him and those around him to their limits.

On June 22, 2012, Tesla invited all of its employees, some select customers, and the press to its factory in Fremont to watch as the first Model S sedans were taken home. Depending on which of the many promised delivery dates you pick, the Model S was anywhere from eighteen months to two-plus years late. Some of the delays were a result of Musk's requests for exotic technologies that needed to be invented. Other delays were simply a function of this still quite young automaker learning how to produce an immaculate luxury vehicle and needing to go through the trial and error tied to becoming a more mature, more refined company.

The outsiders were blown away by their first glimpse of the Tesla factory. Musk had T-E-S-L-A painted in enormous black letters on the side of the building so that people driving by on the freeway, or flying above for that matter, were made well aware of the company's presence. The inside of the factory, once dressed in the dark, dingy tones of General Motors and Toyota, had taken on the Musk aesthetic. The floors received a white epoxy, the walls and beams were painted white, the thirty-foot tall stamping machines were white, and then much of the other machinery, like the teams of the robots, had been painted red, making the place look like an industrial version of Santa Claus's workshop. Just as he did at SpaceX, Musk placed the desks of his engineers right on the factory floor, where they worked in an area cordoned off by rudimentary cubicle dividers. Musk had a desk in this area as well.*

The Model S launch event took place in a section of the factory where they finish off the cars. There's a part of the floor with various grooves and bumps that the cars pass over, as technicians listen for any rattles. There's also a chamber where water can be sprayed at high pressure onto the car to check for leaks. For the very last inspection, the Model S cruises onto a raised platform made out of bamboo, which, when coupled with lots of LED lighting, is meant to provide an abundant amount of contrast so that people can spot flaws on the body. For the few first months that the Model S came off the line, Musk went to this bamboo stage to inspect every vehicle. "He was down on all fours looking up under the wheel well," said Steve Jurvetson, the investor and Tesla board member.

Hundreds of people had gathered around this stage to watch as the first dozen or so cars were presented to their owners. Many of the employees were factory workers who had once been part of the autoworkers' union, lost their jobs when the NUMMI plant closed, and were now back at work again, making the car of the future. They waved American flags and wore red, white, and blue visors. A handful of the workers cried as the Model S sedans were lined up on the stage. Even Musk's most cynical critics would have softened for a moment while watching the proceedings. Say what you will about Tesla receiving government money or hyping up the promise of the electric

---

* "He picks the most visible place on purpose," said the investor and Tesla board member Steve Jurvetson. "He's at Tesla just about every Saturday and Sunday and wants people to see him and know they can find him. Then, he can also call suppliers on the weekend, and let them know that he's personally putting in the hours on the factory floor and expects the same from them."

car, it was trying to do something big and different, and people were getting hired by the thousands as a result. With machines humming in the background, Musk gave a brief speech and then handed the owners their keys. They drove off the bamboo platform and out the factory doors, while the Tesla employees provided a standing ovation.

Just four weeks earlier, SpaceX had flown cargo to the International Space Station and had its capsule returned to Earth—firsts all around for a private company. That feat coupled with the launch of the Model S led to a rapid transformation in the way the world outside of Silicon Valley perceived Musk. The guy who was always promising, promising, promising was doing—and doing spectacular things. "I may have been optimistic with respect to the timing on some of these things, but I didn't overpromise on the outcome," Musk told me during an interview after the Model S launch. "I have done everything I said I was going to do."

Musk did not have Riley around to celebrate with and share in this run of good fortune. They had divorced, and Musk had begun to think about dating again, if he could find the time. Even with this turmoil in his personal life, however, Musk had reached a point of calm that he had not felt in many years. "My main emotion is that there is a bit of weight off my shoulders," he said at the time. Musk took his boys to Maui to meet up with Kimbal and other relatives, marking his first real vacation in a number of years.

It was right after this holiday that Musk let me have the first substantial glimpse into his life. Skin still peeling off his sunburnt arms, Musk met with me at the Tesla and SpaceX headquarters, at the Tesla design studio, and at a Beverley Hills screening of a documentary he had helped sponsor. The film, *Baseball in the Time of Cholera*, was good but grim and explored a cholera outbreak in Haiti. It turned out that Musk had visited Haiti the previous Christmas, filling his jet with toys and MacBook Airs for an orphanage. Bryn Mooser, the codirector of the film, told me that during a barbecue Musk had taught the kids how to fire off model rockets and then later went to visit a village deeper in the jungle by traveling in a dugout canoe. After the screening, Musk and I hung out on the street for a bit away from the crowd. I noted aloud that everyone wants to make him out as the Tony Stark character but that he didn't really exude that "playboy drinking scotch while zooming through Afghanistan in an army convoy" vibe. He fired back, pointing to the Haitian canoe ride. "I got wasted, too, on some drink they call the Zombie," Musk said. He smiled and then invited me to grab some drinks across the street at Mr. Chow to celebrate the movie. All seemed to be going well for Musk, and he savored the moment.

This restful period did not last long and soon enough Tesla's battle for survival resumed. The company could only produce about ten sedans per week at the outset and had thousands of back orders that it needed to fulfill. Short sellers, those investors who bet a company's share price will fall, had taken huge positions in Tesla, making it the most shorted stock out of one hundred of the largest companies listed on the NASDAQ exchange. The naysayers expected numerous Model S flaws to crop up and undermine the enthusiasm for the car, to the point that people started canceling their orders in bulk. There were also huge doubts that Tesla could ramp up production in a meaningful way and do so profitably. In October 2012, the presidential hopeful Mitt Romney dubbed Tesla "a loser," while slagging off a couple of other government-backed green technology companies (the solar panel maker Solyndra and Fisker) during a debate with Barack Obama.[14]

While the doubters placed huge wagers on Tesla's impending

failure, Musk's bluster mode engaged. He began talking about Tesla's goals to become the most profitable major automobile maker in the world, with better margins than BMW. Then, in September 2012, he unveiled something that shocked both Tesla critics and proponents alike. Tesla had secretly been building the first leg of a network of charging stations. The company disclosed the location of six stations in California, Nevada, and Arizona and promised that hundreds more would be on the way. Tesla intended to build a global charging network that would let Model S owners making long drives pull off the highway and recharge very quickly. And they would be able to do so for free. In fact, Musk insisted that Tesla owners would soon be able to travel across the United States without spending a penny on fuel. Model S drivers would have no trouble finding these stations, not only because the cars' onboard computers would guide them to the nearest one but because Musk and von Holzhausen had designed giant red and white monoliths to herald the appearance of the stations.

The Supercharging stations, as Tesla called them, represented a huge investment for the strapped company. An argument could easily be made that spending money on this sort of thing at such a precarious moment in the Model S and Tesla's history was somewhere between daft and batshit crazy. Surely Musk did not have the gall to try to revamp the very idea of the automobile and build an energy network at the same time with a budget equivalent to what Ford and ExxonMobil spend on their annual holiday parties. But that was the exact plan. Musk, Straubel, and others inside Tesla had mapped out this all-or-nothing play long ago and built certain features into the Model S with the Superchargers in mind.*

---

* Tesla got its start using the same lithium ion batteries that go into consumer electronics like laptops. During the early days of the Roadster, this proved a risky but calculated choice. Tesla wanted to tap into Asia's mature battery suppliers and get access to cheap products that

While the arrival of the Model S and the charging network garnered Tesla a ton of headlines, it remained unclear if the positive press and good vibes would last. Serious trade-offs had been made as Tesla rushed to get the Model S to market. The car had some spectacular, novel features. But everyone inside of the company knew that as far as luxury sedans went, the Model S did not match up feature to feature with cars from BMW and Mercedes-Benz. The first few thousand Model S cars, for example, would

---

would keep improving over time. The press played up Tesla's use of these types of batteries, and consumers were fascinated by the idea that a car could be powered by the same energy source sitting inside of their gizmos.

There's a major misconception that Tesla still depends on these types of batteries. Yes, the batteries inside the Model S look like those found in a laptop. Tesla, however, started developing its own battery chemistry in conjunction with partners like Panasonic dating back to late models of the Roadster. Tesla can still use the same manufacturing equipment as consumer electronics companies while ending up with a battery that's safer and better tuned to the intense charging demands of its cars. Along with the secret formula for the battery cells themselves, Tesla has improved the performance of its batteries by developing its own techniques for linking the cells together and cooling them. The battery cells have been designed to vent heat in a very particular way, and there's coolant running throughout the entire battery pack. The battery packs are assembled at the Tesla factory in an area hidden from visitors.

The chemistry, the batteries, the battery pack design—these are all elements of a large, continuous system that Tesla has built from the ground up to allow its cars to charge at record speed. To control the heat produced during the charging process, Tesla has designed an interlinked system of radiators and chillers to cool both the batteries and the chargers. "You've got all that hardware plus the software management system and other controllers," said J. B. Straubel. "All of these things are running at maximum rate." A Model S can recharge 150 miles of range in 20 minutes at one of Tesla's charging stations with DC power pumping straight into the batteries. By comparison, a Nissan Leaf that maxes out at 80 miles of range can take 8 hours to recharge.

ship without the parking sensors and radar-assisted cruise control common on other high-end cars. "It was either hire a team of fifty people right away to make one of these things happen or implement things as best and as fast as you could," Javidan said.

The subpar fit and finish also proved hard to explain. The early adopters could tolerate a windshield wiper going haywire for a couple of days, but they wanted to see seats and visors that met the $100,000 price tag. While Tesla did its best to source the highest-quality materials, it struggled at times to convince the top suppliers to take the company seriously.[15] "People were very suspect that we would deliver one thousand Model Ss," said von Holzhausen. "It was frustrating because we had the drive internally to make the car perfect but could not get the same commitment externally. With something like the visor, we ended up having to go to a third-rate supplier and then work on fixing the situation after the car had already started shipping." The cosmetic issues, though, were minor compared to a tumultuous set of internal circumstances, revealed in detail here for the first time, that threatened to bankrupt the company once again.

Musk had hired George Blankenship, a former Apple executive, to run its stores and service-center operations. At Apple, Blankenship worked just a couple of doors down from Steve Jobs and received credit for building much of the Apple Store strategy. When Tesla first hired Blankenship, the press and public were atwitter, anticipating that'd he do something spectacular and at odds with the traditions of the automotive industry.

Blankenship did some of that. He expanded Tesla's number of stores throughout the world and imbued them with that Apple Store vibe. Along with showcasing the Model S, the Tesla stores sold hoodies and hats and had areas in the back where kids would find crayons and Tesla coloring books. Blankenship gave me a tour of the Tesla store on Santana Row, the glitzy shopping cen-

ter in San Jose. He came off as a warm, grandfatherly sort who
saw Tesla as his chance to make a difference. "The typical dealer
wants to sell you a car on the spot to clear inventory off his lot,"
Blankenship said. "The goal here is to develop a relationship with
Tesla and electric vehicles." Tesla, he said, wanted to turn the
Model S into more than a car. Ideally it would be an object of
desire just like the iPod and iPhone. Blankenship noted that Tesla
had more than ten thousand reservations for the Model S at the
time, the vast majority of which had arrived without the customers test-driving the car. A lot of this early interest resulted from
the aura surrounding Musk, who Blankenship said came off as
similar to Jobs but with a toned-down control-freak vibe. "This
is the first place I have worked that is going to change the world,"
Blankenship said, taking a jab at the sometimes trivial nature of
Apple's gadgets.

While Musk and Blankenship got along at first, their relationship fell apart during the latter stages of 2012. Tesla did have
a large number of reservations in which people put down $5,000
for the right to buy a Model S and get in the purchase queue. But
the company had struggled to turn these reservations into actual
sales. The reasons behind this problem remain unclear. It may
have been that the complaints about the interior and the early
kinks mentioned on the Tesla forums and message boards were
causing concerns. Tesla also lacked financing options to soften
the blow of buying a $100,000 car, while uncertainty surrounded
the resale market for the Model S. You might end up with the
car of the future or you might spend six figures on a dud with a
battery pack that loses its capacity, and with no secondary buyer.
Tesla's service centers at the time were also terrible. The early
cars were unreliable and customers were being sent in droves
to centers unprepared to handle the volume. Many prospective
Tesla owners likely wanted to hang out on the sidelines for a bit

longer to make sure that the company would remain viable. As Musk put it, "The word of mouth on the car sucked."

By the middle of February 2013, Tesla had fallen into a crisis state. If it could not convert its reservations to purchases quickly, its factory would sit idle, costing the company vast amounts of money. And if anyone caught wind of the factory slowdown, Tesla's shares would likely plummet, prospective owners would become even more cautious, and the short sellers would win. The severity of this problem had been hidden from Musk, but once he learned about it, he acted in his signature all-or-nothing fashion. Musk pulled people from recruiting, the design studio, engineering, finance, and wherever else he could find them and ordered them to get on the phone, call people with reservations, and close deals. "If we don't deliver these cars, we are fucked," Musk told the employees. "So, I don't care what job you were doing. Your new job is delivering cars." He placed Jerome Guillen, a former Daimler executive, in charge of fixing the service issues. Musk fired senior leaders whom he deemed subpar performers and promoted a flood of junior people who had been doing above-average work. He also made an announcement personally guaranteeing the resale price of the Model S. Customers would be able to resell their cars for the average going rate of similar luxury sedans with Musk putting his billions behind this pledge. And then Musk tried to orchestrate the ultimate fail-safe for Tesla just in case his maneuvers did not work.

During the first week of April, Musk reached out to his friend Larry Page at Google. According to people familiar with their discussion, Musk voiced his concerns about Tesla's ability to survive the next few weeks. Not only were customers failing to convert their reservations to orders at the rate Musk hoped, but existing customers had also started to defer their orders as they heard about upcoming features and new color choices. The situation got so bad that Tesla had to shut down its factory. Publicly, Tesla said it needed to conduct maintenance on the factory, which was technically true, although the company would have soldiered on had the orders been closing as expected. Musk explained all of this to Page and then struck a handshake deal for Google to acquire Tesla.

While Musk did not want to sell, the deal seemed like the only viable course for Tesla's future. Musk's biggest fear about an acquisition was that the new owner would not see Tesla's goals through to their conclusion. He wanted to make sure that the company would end up producing a mass-market electric vehicle. Musk proposed terms under which he would remain in control of Tesla for eight years or until it started pumping out a mass-market car. Musk also asked for access to $5 billion in capital for factory expansions. Some of Google's lawyers were put off by these demands, but Musk and Page continued to talk about the deal. Given Tesla's value at the time, it was thought that Google would need to pay about $6 billion for the company.

As Musk, Page, and Google's lawyers debated the parameters of an acquisition, a miracle happened. The five hundred or so people whom Musk had turned into car salesmen quickly sold a huge volume of cars. Tesla, which only had a couple weeks of cash left in the bank, moved enough cars in the span of about fourteen days to end up with a blowout first fiscal quarter. Tesla stunned Wall Street on May 8, 2013, by posting its first-ever profit as a public company—$11 million—on $562 million in sales. It delivered 4,900 Model S sedans during the period. This announcement sent Tesla's shares soaring from about $30 a share to $130 per share in July. Just a couple of weeks after revealing the first-quarter results, Tesla paid off its $465 million loan from the government early and with interest. Tesla suddenly appeared to have vast cash reserves at its disposal, and the short sellers were forced to take massive losses. The solid performance of the stock increased consumers' confidence, creating a virtuous circle for Tesla. With cars selling and Tesla's value rising, the deal with Google was no longer necessary, and Tesla had become too expensive to buy. The talks with Google ended.*

What transpired next was the Summer of Musk. Musk put his public relations staff on high alert, telling them that he wanted to try to have one Tesla announcement per week. The company never quite lived up to that pace, but it did issue statement after statement. Musk held a series of press conferences that addressed financing for the Model S, the construction of more charging stations, and the opening of more retail stores. During one announcement, Musk noted that Tesla's charging stations were solar-powered and had batteries on-site to store extra juice. "I was joking that even if there's some zombie apocalypse, you'll still be able to travel throughout the country using the Tesla Supercharger system," Musk said, setting the bar very high for CEOs at other automakers. But the biggest event by far was held in Los Angeles, where Tesla unveiled another secret feature of the Model S.

In June 2013, Tesla cleared the prototype vehicles out of its Los Angeles design studio and invited Tesla owners and the media for a flashy evening soiree. Hundreds of people showed up, driving their pricey Model S sedans through the grungy streets of Hawthorne and parking in between the design studio and the SpaceX factory. The studio had been converted into a lounge. The light-

---

* Google's attorneys had asked to make a presentation to Tesla's board. Before he would permit this, Musk asked for the right to call on Google for a loan in case Tesla encountered cash flow issues after acquisition talks became public, as there would otherwise be no way for Tesla to raise money. Google hesitated on this for a few weeks, by which time Tesla ended up in the clear.

ing was dim, and the floor had been covered in AstroTurf and tiered to make plateaus where people could mingle or plop down on couches. Women in tight black dresses cruised through the crowd, serving drinks. Daft Punk's "Get Lucky" played on the sound system. A stage had been built at the front of the room, but before Musk ascended it he mingled with the masses. It was clear that he had become a rock star for Tesla owners—every bit the equivalent of Steve Jobs for the Apple faithful. People surrounded him and asked to take pictures. Meanwhile, Straubel stood off to the side, often totally alone.

After people had a couple of drinks, Musk fought through the crowd to the front of the room, where old TV commercials projected onto a screen above the stage showed families stopping by Esso and Chevron stations. The kids were so happy to see the Esso tiger mascot. "Gas is a weird thing to love," Musk said. "Honestly." That's when he brought a Model S up onstage. A hole opened up in the floor beneath the car. It had been possible all along, Musk said, to replace the battery pack underneath the Model S in a matter of seconds—the company just hadn't told anyone about this. Tesla would now start adding battery swapping at its charging stations as a quicker option to recharging. Someone could drive right over a pit where a robot would take off the car's battery pack and install a new one in ninety seconds, at a cost equivalent to filling up with a tank of gas. "The only decision that you have to make when you come to one of our Tesla stations is do you prefer faster or free," Musk said.

---

* Following the demonstration, Tesla struggled to deliver on the battery swap technology. Musk had promised that the first few stations would arrive in late 2013. A year after the event, though, Tesla had yet to open a single station. According to Musk, the company ended up needing to deal with more pressing issues. "We're going to do it because we said we'd do it," Musk said. "It may not be on the schedule that we'd like but we always come through in the end."

In the months that followed, a couple of events threatened to derail the Summer of Musk. The *New York Times* penned a withering review of the car and its charging stations, and a couple of the Model S sedans caught fire after being involved in collisions. Disobeying conventional public relations wisdom, Musk went after the reporter, using data pulled from the car to undermine the reviewer's claims. Musk penned the feisty rebuttal himself, while on vacation in Aspen with Kimbal, and friend and Tesla board member Antonio Gracias. "At some other company, it would be a public relations group putting something like this together," Gracias said. "Elon felt like it was the most important problem facing Tesla at the time and that's always what he deals with and how he prioritizes. It could kill the car and represented an existential threat against the business. Have there been moments where his unconventional style in these types of situations has made me cringe? Yes. But I trust that it will work out in the end." Musk applied a similar approach to dealing with the fires by declaring the Model S the safest car in America in a press release and adding a titanium underbody shield and aluminum plates to the vehicle to deflect and destroy debris and keep the battery pack safe.[16]

The fires, the occasional bad review—none of this had any effect on Tesla's sales or share price. Musk's star shone brighter and brighter as Tesla's market value ballooned to about half that of GM and Ford.

Tesla held another press event in October 2014 that cemented Musk's place as the new titan of the auto industry. Musk unveiled a supercharged version of the Model S with two motors—one in the front and one in the back. It could go zero to 60 in 3.2 seconds. The company had turned a sedan into a supercar. "It's like taking off from a carrier deck," Musk said. "It's just bananas." Musk also unveiled a new suite of software for the Model S that gave it autopilot functions. The car had radar to detect objects and warn of possible collisions and could guide itself via GPS. "Later, you will be able to summon the car," Musk said. "It will come to wherever you are. There's also something else I would like to do. Many of our engineers will be hearing this in real time. I would like the charge connector to plug itself into the car, sort of like an articulating snake. I think we will probably do something like that."

Thousands of people waited in line for hours to see Musk demonstrate this technology. Musk cracked jokes during the presentation and played off the crowd's enthusiasm. The man who had been awkward in front of media during the PayPal years had developed a unique, slick stagecraft. A woman standing next to me in the crowd went weak in the knees when Musk first took the stage. A man to my other side said he wanted a Model X and had just offered $15,000 to a friend to move up on the reservation list, so that he could end up with model No. 700. The enthusiasm coupled with Musk's ability to generate attention was emblematic of just how far the little automaker and its eccentric CEO had come. Rival car companies would kill to receive such interest and had basically been left dumbfounded as Tesla snuck up on them and delivered more than they had ever imagined possible.

As the Model S fever gripped Silicon Valley, I visited Ford's small research and development lab in Palo Alto. The head of the lab at the time was a ponytailed, sandal-wearing engineer named T. J. Giuli, who felt very jealous of Tesla. Inside of every Ford were dozens of computing systems made by different companies that all had to speak to each other and work as one. It was a mess of complexity that had evolved over time, and simplifying the situation would prove near impossible at this point, especially for a company like Ford, which needed to pump out hundreds of thousands of cars per year and could not afford to stop and reboot.

Tesla, by contrast, got to start from scratch and make its own software the focus of the Model S. Giuli would have loved the same opportunity. "Software is in many ways the heart of the new vehicle experience," he said. "From the powertrain to the warning chimes in the car, you're using software to create an expressive and pleasing environment. The level of integration that the software has into the rest of the Model S is really impressive. Tesla is a benchmark for what we do here." Not long after this chat, Giuli left Ford to become an engineer at a stealth start-up.

There was little the mainstream auto industry could do to slow Tesla down. But that didn't stop executives from trying to be difficult whenever possible. Tesla, for example, wanted to call its third-generation car the Model E, so that its lineup of vehicles would be the Model S, E, and X—another playful Musk gag. But Ford's then CEO, Alan Mulally, blocked Tesla from using Model E, with the threat of a lawsuit. "So I call up Mulally and I was like, 'Alan, are you just fucking with us or are you really going to do a Model E?'" Musk said. "And I'm not sure which is worse. You know? Like it would actually make more sense if they're just fucking with us because if they actually come out with a Model E at this point, and we've got the Model S and the X and Ford comes out with the Model E, it's going to look ridiculous. So even though Ford did the Model T a hundred years ago, nobody thinks of 'Model' as being a Ford thing anymore. So it would just feel like they stole it. Like why did you go steal Tesla's E? Like you're some sort of fascist army marching across the alphabet, some sort of *Sesame Street* robber. And he was like, 'No, no, we're definitely going to use it.' And I was like, 'Oh, I don't think that's such a good idea because people are going to be confused because it's not going to make sense. People aren't used to Ford having Model something these days. It's usually called like the Ford Fusion.' And he was like, no, his guys really want to use that. That's terrible." After that, Tesla registered the trademark for Model Y as another joke. "In fact, Ford called us up deadpan and said, 'We see you've registered Model Y. Is that what you're going to use instead of the Model E?'" Musk said. "I'm like, 'No, it's a joke. S-E-X-Y. What does that spell?' But trademark law is a dry profession it turns out."*

What Musk had done that the rival automakers missed or didn't have the means to combat was turn Tesla into a lifestyle. It did not just sell someone a car. It sold them an image, a feeling they were tapping into the future, a relationship. Apple did the same thing decades ago with the Mac and then again with the iPod and iPhone. Even those who were not religious about their affiliation to Apple were sucked into its universe once they bought the hardware and downloaded software like iTunes.

This sort of relationship is hard to pull off if you don't control as much of the lifestyle as possible. PC makers that farmed their software out to Microsoft, their chips to Intel, and their design to Asia could never make machines as beautiful and as complete as Apple's. They also could not respond in time as Apple took this expertise to new areas and hooked people on its applications.

You can see Musk's embrace of the car as lifestyle in Tesla's abandonment of model years. Tesla does not designate cars as being 2014s or 2015s, and it also doesn't have "all the 2014s in

---

* As for the origins of the Model S name, Musk said, "Well, I like calling things what they are. We had the Roadster, but there was no good word for a sedan. You can't call it the Tesla Sedan. That's boring as hell. In the U.K., they say 'saloon,' but then it's sort of like, 'What are you? A cowboy or something?' We went through a bunch of iterations, and the Model S sounded the best. And it was like a vague nod to Ford being the Model T in that electric cars preceded the Model T, and in a way we're coming full circle and the thing that proceeded the Model T is now going into production in the twenty-first century, hence the Model S. But that's sort of more like reversing the logic."

stock must go, go, go and make room for the new cars" sales. It produces the best Model S it can at the time, and that's what the customer receives. This means that Tesla does not develop and hold on to a bunch of new features over the course of the year and then unleash them in a new model all at once. It adds features one by one to the manufacturing line when they're ready. Some customers may be frustrated to miss out on a feature here and there. Tesla, however, manages to deliver most of the upgrades as software updates that everyone gets, providing current Model S owners with pleasant surprises.

For the Model S owner, the all-electric lifestyle translates into a less hassled existence. Instead of going to the gas station, you just plug the car in at night, a rhythm familiar to anyone with a smartphone. The car will start charging right away or the owner can tap into the Model S's software and schedule charging to take place late at night, when the cheapest electricity rates are available. Tesla owners not only dodge gas stations; they mostly get to skip out on visits to mechanics. A traditional vehicle needs oil and transmission fluid changes to deal with all the friction and wear and tear produced by its thousands of moving parts. The simpler electric car design eliminates this type of mainte- nance. Both the Roadster and the Model S also take advantage of what's known as regenerative braking, which extends the life of the brakes. During stop-and-go situations, the Tesla will brake by kicking the motor into reverse via software and slowing down the wheels instead of using brake pads and friction to clamp them down. The Tesla motor generates electricity during this process and funnels it back to the batteries, which is why electric cars get better mileage in city traffic. Tesla still recommends that owners bring in the Model S once a year for a checkup but that's mostly to give the vehicle a once-over and make sure that none of the components seems to be wearing down prematurely.

Even Tesla's approach to maintenance is philosophically different from that of the traditional automotive industry. Most car dealers make the majority of their profits from servicing cars. They treat vehicles like a subscription service, expecting people to visit their service centers multiple times a year for many years. This is the main reason dealerships have fought to block Tesla from selling its cars directly to consumers.* "The ultimate goal is to never have to bring your car back in after you buy it," said Javidan. The dealers charge more than independent mechanics but give people the peace of mind that their car is being worked on by a specialist for a particular make of vehicle. Tesla makes its profits off the initial sale of the car and then from some optional software services. "I got the number ten Model S," said Konstantin Othmer,[17] the Silicon Valley software whiz and entrepreneur. "It was an awesome car, but it had just about every issue you might have read about in the forums. They would fix all these things and decided to trailer the car back to the shop so that they didn't add any miles to it. Then I went in for a one-year service, and they spruced up everything so that the car was better than new. It was surrounded by velvet ropes in the service center. It was just beautiful."

Tesla's model isn't just about being an affront to the way carmakers and dealers do business. It's a more subtle play on how electric cars represent a new way to think of automobiles. All car companies will soon follow Tesla's lead and offer some form of

---

* A handful of lawsuits have been filed against Tesla with auto dealers arguing that the company should not be able to sell its cars directly. But even in those states that have banned Tesla's stores, prospective customers can usually request a test drive, and someone from Tesla will show up with a vehicle. "Sometimes you have to put something out there for people to attack," Musk said. "In the long run, the stores won't be important. The way things will really grow is by word of mouth. The stores are like a viral seed to get things going."

over-the-air updates to their vehicles. The practicality and scope
of their updates will be limited, however. "You just can't do an
over-the-air sparkplug change or replacement of the timing belt,"
said Javidan. "With a gas car, you have to get under the hood at
some point and that forces you back to the dealership anyway.
There's no real incentive for Mercedes to say, 'You don't need to
bring the car in,' because it's not true." Tesla also has the edge of
having designed so many of the key components for its cars in-
house, including the software running throughout the vehicle. "If
Daimler wants to change the way a gauge looks, it has to contact a
supplier half a world away and then wait for a series of approvals,"
Javidan said. "It would take them a year to change the way the 'P'
on the instrument panel looks. At Tesla, if Elon decides he wants
a picture of a bunny rabbit on every gauge for Easter, he can have
that done in a couple of hours."*

As Tesla turned into a star of modern American industry, its
closest rivals were obliterated. Fisker Automotive filed for bankruptcy and was bought by a Chinese auto parts company in 2014.
One of its main investors was Ray Lane, a venture capitalist at
Kleiner Perkins Caufield & Byers. Lane had cost Kleiner Perkins
a chance to invest in Tesla and then backed Fisker—a disastrous
move that tarnished the firm's brand and Lane's reputation. Bet-

---

* Or as Straubel put it, "Watching people drive the Model S across the
  country is phenomenal. There is no way you can do that in anything
  else. It's not about putting a charging station in the desert as a stunt.
  It's about realizing where this is going to go. We will end up launching the third-generation car into a world where this charging network
  is free and ubiquitous. It bugs me when people compare us to a car
  company. The cars are absolutely our main product, but we are also
  an energy company and a technology company. We are going down
  to the dirt and having discussions with mining companies about the
  materials for our batteries and going up to commercialize all the
  pieces that make up an electronic vehicle and all the pieces that make
  an awesome product."

11

THE UNIFIED FIELD THEORY
OF ELON MUSK

THE RIVE BROTHERS USED TO BE LIKE A TECHNOLOGY GANG. IN THE LATE 1990s, they would jump on skateboards and zip around the streets of Santa Cruz, knocking on the doors of businesses and asking if they needed any help managing their computing systems. The young men, who had all grown up in South Africa with their cousin Elon Musk, soon decided there must be an easier way to hawk their technology smarts than going door-to-door. They wrote some software that allowed them to take control of their clients' systems from afar and to automate many of the standard tasks that companies required, such as installing updates for applications. The software became the basis of a new company called Everdream, and the brothers promoted their technology in some compelling ways. Billboards went up around Silicon Valley in which Lyndon Rive, a buff underwater hockey player,* stood naked with his pants around his ankles, while holding a computer in front of his crotch. Up above his photo, the tagline for the ad read, "Don't get caught with your systems down."

By 2004, Lyndon and his brothers, Peter and Russ, wanted a new challenge—something that not only made them money but, as Lyndon put it, "something that made us feel good every single day." Near the end of the summer that year, Lyndon rented an RV and set out with Musk for the Black Rock desert and the madness of Burning Man. The men used to go on adventures all the time when they were kids and looked forward to the long drive as a way to catch up and brainstorm about their businesses. Musk knew that Lyndon and his brothers were angling for something big. While driving, Musk turned to Lyndon and suggested that he look into the solar energy market. Musk had studied it a bit and thought there were some opportunities that others had missed. "He said it was a good place to get into," Lyndon recalled.

After arriving at Burning Man, Musk, a regular at the event, and his family went through their standard routines. They set up camp and prepped their art car for a drive. This year, they had cut the roof off a small car, elevated the steering wheel, shifted it to the right so that it was placed near the middle of the vehicle, and replaced the seats with a couch. Musk took a lot of pleasure in driving the funky creation.[19] "Elon likes to see the rawness of people there," said Bill Lee, his longtime friend. "It's his version of camping. He wants to go and drive the art cars and see installations and the great light shows. He dances a lot." Musk put on a display of strength and determination at the event as well. There

---

* No, really. Both Lyndon and his wife play underwater hockey and used these skills to secure green cards, meeting the criteria for the "exceptional abilities" the United States desires. They ultimately played for the U.S. national teams.

was a wooden pole perhaps thirty feet high with a dancing platform at the top. Dozens of people tried and failed to climb it, and
then Musk gave it a go. "His technique was very awkward, and he
should not have succeeded," said Lyndon. "But he hugged it and
just inched up and inched up until he reached the top."

Musk and the Rives left Burning Man enthused. The Rives
decided to become experts on the solar industry and find the
opportunity in the market. They spent two years studying solar
technology and the dynamics of the business, reading research
reports, interviewing people, and attending conferences along
the way. It was during the Solar Power International conference
that the Rive brothers really hit on what their business model
might be. Only about two thousand* people showed up for the
event, and they all fit into a couple of hotel conference rooms
for presentations and panels. During one open discussion session,
representatives from a handful of the world's largest solar install-
ers were sitting onstage, and the moderator asked what they were
doing to make solar panels more affordable for consumers. "They
all gave the same answer," Lyndon said. "They said, 'We're waiting for the cost of the panels to drop.' None of them were taking
ownership of the problem."

At the time, it was not easy for consumers to get solar panels
on their houses. You had to be very proactive, acquiring the panels and finding someone else to install them. The consumer paid
up front and had to make an educated guess as to whether or not
his or her house even got enough sunshine to make the ordeal
worthwhile. On top of all this, people were reluctant to buy panels, knowing that the next year's models would be more efficient.

The Rives decided to make buying into the solar proposition
much simpler and formed a company called SolarCity in 2006.

---

* Thirteen thousand people showed up in 2013.

Unlike other companies, they would not manufacture their own solar panels. Instead they would buy them and then do just about everything else in-house. They built software for analyzing a customer's current energy bill and the position of their house and the amount of sunlight it typically received to determine if solar made sense for the property. They built up their own teams to install the solar panels. And they created a financing system in which the customer did not need to pay anything up front for the panels. The consumer leased the panels over a number of years at a fixed monthly rate. Consumers got a lower bill overall, they were no longer subject to the constantly rising rates of typical utilities, and, if they sold their house, they could pass the contract to the new owner. At the end of the lease, the homeowner could also upgrade to new, more efficient panels. Musk had helped his cousins come up with this structure and become the company's chairman and its largest shareholder, owning about a third of SolarCity.

Six years later, SolarCity had become the largest installer of solar panels in the country. The company had lived up to its initial goals and made installing the panels painless. Rivals were rushing to mimic its business model. SolarCity had benefited along the way from a collapse in the price of solar panels, which occurred after Chinese panel manufacturers flooded the market with product. It had also expanded its business from consumers to businesses with companies like Intel, Walgreens, and Wal-Mart signing up for large installations. In 2012, SolarCity went public and its shares soared higher in the months that followed. By 2014, SolarCity was valued at close to $7 billion.

During the entire period of SolarCity's growth, Silicon Valley had dumped huge amounts of money into green technology companies with mostly disastrous results. There were the automotive flubs like Fisker and Better Place, and Solyndra, the solar cell maker that conservatives loved to hold up as a cautionary tale of government spending and cronyism run amok. Some of the most famous venture capitalists in history, like John Doerr and Vinod Khosla, were ripped apart by the local and national press for their failed green investments. The story was almost always the same. People had thrown money at green technology because it seemed like the right thing to do, not because it made business sense. From new kinds of energy storage systems to electric cars and solar panels, the technology never quite lived up to its billing and required too much government funding and too many incentives to create a viable market. Much of this criticism was fair. It's just that there was this Elon Musk guy hanging around who seemed to have figured something out that everyone else had missed. "We had a blanket rule against investing in clean-tech companies for about a decade," said Peter Thiel, the PayPal cofounder and venture capitalist at Founders Fund. "On the macro level, we were right because clean tech as a sector was quite bad. But on the micro level, it looks like Elon has the two most successful clean-tech companies in the U.S. We would rather explain his success as being a fluke. There's the whole *Iron Man* thing in which he's presented as a cartoonish businessman— this very unusual animal at the zoo. But there is now a degree to which you have to ask whether his success is an indictment on the rest of us who have been working on much more incremental things. To the extent that the world still doubts Elon, I think it's a reflection on the insanity of the world and not on the supposed insanity of Elon."

SolarCity, like the rest of Musk's ventures, did not represent a business opportunity so much as it represented a worldview. Musk had decided long ago—in his very rational manner—that solar made sense. Enough solar energy hits the Earth's surface in about an hour to equal a year's worth of worldwide energy consumption from all sources put together.[20] Improvements in the efficiency of solar panels have been happening at a steady clip. If solar is destined to be mankind's preferred energy source in the future, then this future ought to be brought about as quickly as possible.

Starting in 2014, SolarCity began to make the full extent of its ambitions more obvious. First, the company began selling energy storage systems. These units were built through a partnership with Tesla Motors. Battery packs were manufactured at the Tesla factory and stacked inside refrigerator-sized metal cases. Businesses and consumers could purchase these storage systems to augment their solar panel arrays. Once they were charged up, the battery units could be used to help large customers get through the night or during unexpected outages. Customers could also pull from the batteries instead of the grid during peak energy use periods, when utilities tend to tack on extra charges. While SolarCity rolled the storage units out in a modest, experimental fashion, the company expects most of its customers to buy the systems in the years ahead to smooth out the solar experience and help people and businesses leave the electrical grid altogether.

Then, in June 2014, SolarCity acquired a solar cell maker called Silevo for $200 million. This deal marked a huge shift in strategy. SolarCity would no longer buy its solar panels. It would make them at a factory in New York State. Silevo's cells were said to be 18.5 percent efficient at turning light into energy, compared to 14.5 percent for most cells, and the expectations were that the company could reach 24 percent efficiency with the right manufacturing techniques. Buying, rather than manufacturing, solar panels had been one of SolarCity's great advantages. It could capitalize on the glut in the solar cell market and avoid the large capital expenditures tied to building and running factories. With 110,000 customers, however, SolarCity had started to consume so many solar panels that it needed to ensure a consistent supply and price. "We are currently installing more solar than most of the companies are manufacturing," said Peter Rive, the cofounder and chief technology officer at SolarCity. "If we do the manufacturing ourselves and take advantage of some different technology, our costs will be lower—and this business has always been about lowering the costs."

After adding the leases, the storage units, and the solar cell manufacturing together, it became clear to close observers of SolarCity that the company had morphed into something resembling a utility. It had built out a network of solar systems all under its control and managed by the company's software. By the end of 2015, SolarCity expects to have installed 2 gigawatts' worth of solar panels, producing 2.8 terawatt-hours of electricity per year. "This would put us on a path to fulfill our goal to become one of the largest suppliers of electricity in the United States," the company said after announcing these figures in a quarterly earnings statement. The reality is that SolarCity accounts for a tiny fraction of the United States' annual energy consumption and has a long way to go to become a major supplier of electricity in the country. There can, however, be little doubt that Musk intends for the company to be a dominant force in the solar industry and in the energy industry overall.

What's more, SolarCity is a key part of what can be thought of as the unified field theory of Musk. Each one of his businesses is interconnected in the short term and the long term. Tesla makes battery packs that SolarCity can then sell to end customers. Solar-City supplies Tesla's charging stations with solar panels, helping Tesla to provide free recharging to its drivers. Newly minted Model S owners regularly opt to begin living the Musk Lifestyle and outfit their homes with solar panels. Tesla and SpaceX help each other as well. They exchange knowledge around materials,

manufacturing techniques, and the intricacies of operating factories that build so much stuff from the ground up.

For most of their histories, SolarCity, Tesla, and SpaceX have
been the clear underdogs in their respective markets and gone to
war against deep-pocketed, entrenched competitors. The solar,
automotive, and aerospace industries remain larded down by reg-
ulation and bureaucracy, which favors incumbents. To people in
these industries Musk came off as a wide-eyed technologist who
could be easily dismissed and ridiculed and who, as a competitor,
fell somewhere on the spectrum between annoying and full of
shit. The incumbents did their usual thing using their connections in Washington to make life as miserable as possible on all
three of Musk's companies, and they were pretty good at it.

As of 2012, Musk Co. turned into a real threat, and it became
harder to go at SolarCity, Tesla, or SpaceX as individual companies. Musk's star power had surged and washed over all three
ventures at the same time. When Tesla's shares jumped, quite
often SolarCity's did, too. Similar optimistic feelings accompanied successful SpaceX launches. They proved Musk knew how
to accomplish the most difficult of things, and investors seemed
to buy in more to the risks Musk took with his other enterprises.
The executives and lobbyists of aerospace, energy, and automotive companies were suddenly going up against a rising star of big
business—an industrialist celebrity. Some of Musk's opponents
started to fear being on the wrong side of history or at least the
wrong side of his glow. Others began playing really dirty.

Musk has spent years buttering up the Democrats. He's visited the White House several times and has the ear of President
Obama. Musk, however, is not a blind loyalist. He first and foremost backs the beliefs behind Musk Co. and then uses any prag-
matic means at his disposal to advance his cause. Musk plays the
part of the ruthless industrialist with a fierce capitalist streak bet-

ter than most Republicans and has the credentials to back it up and earn support. The politicians in states like Alabama looking to protect some factory jobs for Lockheed or in New Jersey trying to help out the automobile dealership lobby now have to contend with a guy who has an employment and manufacturing empire spread across the entire United States. As of this writing, SpaceX had a factory in Los Angeles, a rocket test facility in central Texas, and had just started construction on a spaceport in South Texas. (SpaceX does a lot of business at existing launch sites in California and Florida, as well.) Tesla had its car factory in Silicon Valley, the design center in Los Angeles, and had started construction on a battery factory in Nevada. (Politicians from Nevada, Texas, California, New Mexico, and Arizona threw themselves at Musk over the battery factory, with Nevada ultimately winning the business by offering Tesla $1.4 billion in incentives. This event confirmed not only Musk's soaring celebrity but also his unmatched ability to raise funds.) SolarCity has created thousands of white- and blue-collar clean-tech jobs, and it will create manufacturing jobs at the solar panel factory that's being built in Buffalo, New York. All together, Musk Co. employed about fifteen thousand people at the end of 2014. Far from stopping there, the plan for Musk Co. calls for tens of thousands of more jobs to be created on the back of ever more ambitious products.

Tesla's primary focus throughout 2015 will be bringing the Model X to market. Musk expects the SUV to sell at least as well as the Model S and wants Tesla's factories to be capable of making 100,000 cars per year by the end of 2015 to keep up with demand for both vehicles. The major downside accompanying the Model X is its price. The SUV will start at the same lofty prices as the Model S, which limits the potential customer base. The hope, though, is that the Model X turns into the luxury vehicle of choice for families and solidifies the Tesla brand's connection with women. Musk has pledged that the Supercharger network, service centers, and the battery-swap stations will be built out even more in 2015 to greet the arrival of the new vehicle. Beyond the Model X, Tesla has started work on the second version of the Roadster, talked about making a truck, and, in all seriousness, has begun modeling a type of submarine car that could transition from road to water. Musk paid $1 million for the Lotus Esprit that Roger Moore drove underwater in *The Spy Who Loved Me* and wants to prove that such a vehicle can be done. "Maybe we'll make two or three, but it wouldn't be more than that," Musk told the *Independent* newspaper. "I think the market for submarine cars is quite small."

At the opposite end of the sales spectrum, or so Musk hopes, will be Tesla's third-generation car, or the Model 3. Due out in 2017, this four-door car would come in around $35,000 and be the real measure of Tesla's impact on the world. The company hopes to sell hundreds of thousands of the Model 3 and make electric cars truly mainstream. For comparison, BMW sells about 300,000 Minis and 500,000 of its BMW 3 Series vehicles per year. Tesla would look to match those figures. "I think Tesla is going to make a lot of cars," Musk said. "If we continue on the current growth rate, I think Tesla will be one of the most valuable companies in the world."

Tesla already consumes a huge portion of the world's lithium ion battery supply and will need far more batteries to produce the Model 3. This is why, in 2014, Musk announced plans to build what he dubbed the Gigafactory, or the world's largest lithium ion manufacturing facility. Each Gigafactory will employ about 6,500 people and help Tesla meet a variety of goals. It should first allow Tesla to keep up with the battery demand created by its cars and the storage units sold by SolarCity. Tesla also expects to be able to lower the costs of its batteries while improving their energy density. It will build the Gigafactory in conjunction with longtime battery partner Panasonic, but it will be Tesla that is running the factory and fine-tuning its operations. According to Straubel, the battery packs coming out of the Gigafactory should be dramatically cheaper and better than the ones built today, allowing Tesla not only to hit the $35,000 price target for the Model 3 but also to pave the way for electric vehicles with 500-plus miles of range.

If Tesla actually can deliver an affordable car with 500 miles of range, it will have built what many people in the auto industry insisted for years was impossible. To do that while also constructing a worldwide network of free charging stations, revamping the way cars are sold, and revolutionizing automotive technology would be an exceptional feat in the history of capitalism.

In early 2014, Tesla raised $2 billion by selling bonds. Tesla's ability to raise money from eager investors was a newfound luxury. Tesla had bordered on bankruptcy for much of its existence and been one major technical gaffe from obsolescence at all times. The money coupled with Tesla's still-rising share price and strong sales has put the company in a position to open lots of new stores and service centers while advancing its manufacturing capabilities. "We don't necessarily need all of the money for the Gigafactory right now, but I decided to raise it in advance because you never know when there will be some bloody meltdown," Musk said. "There could be external factors or there could be some unexpected recall and then suddenly we need to raise money on top of dealing with that. I feel a bit like my grandmother. She lived through the Great Depression and some real hard times. Once you've been through that, it stays with you for a long time. I'm not sure it ever leaves really. So, I do feel joy now, but there's still that nagging feeling that it might all go away.

Even later in life when my grandmother knew there was really no possibility of her going hungry, she always had this thing about food. With Tesla, I decided to raise a huge amount of money just in case something terrible happens."

Musk felt optimistic enough about Tesla's future to talk to me about some of his more whimsical plans. He hopes to redesign the Tesla headquarters in Palo Alto, a change employees would welcome. The building, with its tiny, 1980s-era lobby and a kitchen that can barely handle a few people making cereal[21] at the same time, has none of the perks of a typical Silicon Valley darling. "I think our Tesla headquarters looks like crap," Musk said. "We're going to spruce things up. Not to sort of the Google level. You have to be like making money hand over fist in order to be able to spend money the way that Google does. But we're going to make our headquarters much nicer and put in a restaurant." Naturally, Musk had ideas for some mechanical enhancements as well. "Everybody around here has slides in their lobbies," he said. "I'm actually wondering about putting in a roller coaster—like a functional roller coaster at the factory in Fremont. You'd get in, and it would take you around factory but also up and down. Who else has a roller coaster? I'm thinking about doing that with SpaceX, too. That one might be even bigger since SpaceX has like ten buildings now. It would probably be really expensive, but I like the idea of it."

What's fascinating is that Musk remains willing to lose it all. He doesn't want to build just one Gigafactory but several. And he needs these facilities to be built quickly and flawlessly, so that they're cranking out massive quantities of batteries right as the Model 3 arrives. If need be, Musk will build a second Gigafactory to compete with the Nevada site and place his own employees in competition with each other in a race to make the batteries first. "We're not really trying to sort of yank anyone's chain here,"

Musk said. "It's just like this thing needs to be completed on time. If we suddenly find that we're leveling the ground and laying the foundation and we're on a bloody Indian burial ground, then fuck. We can't say, 'Oh shit. Let's go back to the other place that we were thinking about and get a six-month reset.' Six months for this factory is a huge deal. Do the basic math and it's more than a billion dollars a month in lost revenue,* assuming we use it to capacity. From a different standpoint, if we spend all the money to prepare the car factory in Fremont to triple the volume from 150,000 per year to 450,000 or 500,000 cars and hire and train all the people, and we're just sitting there waiting for the factory to come on line, we'd be burning money like it was going out of fashion. I think that could kill the company.

"A six-month offset would be like, like Gallipoli. You have to make sure you charge right after the bombardment. Don't fucking sit around for two hours so that the Turks can go back in the trenches. Timing is important. We have to do everything we can to minimize the timing risk."

What Musk struggles to fathom is why other automakers with deeper pockets aren't making similar moves. At a minimum, Tesla seems to have influenced consumers and the auto industry enough for there to be an expected surge in demand for electric vehicles. "I think we have moved the needle for almost every car company," Musk said. "Just the twenty-two thousand cars we sold in 2013 had a highly leveraged effect in pushing the industry toward sustainable technology." It's true that the supply for lithium ion batteries is already constrained, and Tesla looks like the only company addressing the problem in a meaningful way.

---

* If you assume an average selling price of $40,000 per car for 300,000 cars sold in a year, that's $12 billion in annual revenue, or $1 billion per month.

"The competitors are all sort of pooh-poohing the Gigafactory," Musk said. "They think it's a stupid idea, that the battery supplier should just go build something like that. But I know all the suppliers, and I can tell you that they don't like the idea of spending several billion dollars on a battery factory. You've got a chicken-and-egg problem where the car companies are not going to commit to a giant volume because they're not sure you can sell enough electric cars. So, I know we can't get enough lithium ion batteries unless we build this bloody factory, and I know no one else is building this thing."

There's the potential that Tesla is setting itself up to capitalize on a situation like the one Apple found itself in when it first introduced the iPhone. Apple's rivals spent the initial year after the iPhone's release dismissing the product. Once it became clear Apple had a hit, the competitors had to catch up. Even with the device right in their hands, it took companies like HTC and Samsung years to produce anything comparable. Other once-great companies like Nokia and BlackBerry didn't withstand the shock. If, and it's a big if, Tesla's Model 3 turned into a massive hit—the thing that everyone with enough money wanted because buying something else would just be paying for the past—then the rival automakers would be in a terrible bind. Most of the car companies dabbling in electric vehicles continue to buy bulky, off-the-shelf batteries rather than developing their own technology. No matter how much they wanted to respond to the Model 3, the automakers would need years to come up with a real challenger and even then they might not have a ready supply of batteries for their vehicles.

"I think it is going to be a bit like that," Musk said. "When will the first non-Tesla Gigafactory get built? Probably no sooner than six years from now. The big car companies are so derivative. They want to see it work somewhere else before they will approve the project and move forward. They're probably more like seven years away. But I hope I'm wrong."

Musk speaks about the cars, solar panels, and batteries with such passion that it's easy to forget they are more or less sideline projects. He believes in the technologies to the extent that he thinks they're the right things to pursue for the betterment of mankind. They've also brought him fame and fortune. Musk's ultimate goal, though, remains turning humans into an interplanetary species. This may sound silly to some, but there can be no doubt that this is Musk's raison d'être. Musk has decided that man's survival depends on setting up another colony on another planet and that he should dedicate his life to making this happen.

Musk is now quite rich on paper. He was worth about $10 billion at the time of this writing. When he started SpaceX more than a decade ago, however, he had far less capital at his disposal. He didn't have the fuck-you money of a Jeff Bezos, who handed his space company Blue Origin a kingly pile of cash and asked it to make Bezos's dreams come true. If Musk wanted to get to Mars, he would have to earn it by building SpaceX into a real business. This all seems to have worked in Musk's favor. SpaceX has learned to make cheap and effective rockets and to push the limits of aerospace technology.

In the near term, SpaceX will begin testing its ability to take people into space. It wants to perform a manned test flight by 2016 and to fly astronauts to the International Space Station for NASA the next year. The company will also likely make a major move into building and selling satellites, which would mark an expansion into one of the most lucrative parts of the aerospace business. Along with these efforts, SpaceX has been testing the Falcon Heavy—its giant rocket capable of flying the biggest payloads in the world—and its reusable-rocket technology. In early 2015, SpaceX almost managed to land the first stage of its rocket on a platform in the ocean. Once it succeeds, it will begin performing tests on land.

In 2014, SpaceX also began construction on its own spaceport in South Texas. It has acquired dozens of acres where it plans to construct a modern rocket launch facility unlike anything the world has seen. Musk wants to automate a great deal of the launch process, so that the rockets can be refueled, stood up, and fired on their own with computers handling the safety procedures. SpaceX wants to fly rockets several times a month for its business, and having its own spaceport should help speed up such capabilities. Getting to Mars will require an even more impressive set of skills and technology.

"We need to figure out how to launch multiple times a day," Musk said. "The thing that's important in the long run is establishing a self-sustaining base on Mars. In order for that to work—in order to have a self-sustaining city on Mars—there would need to be millions of tons of equipment and probably millions of people. So how many launches is that? Well, if you send up 100 people at a time, which is a lot to go on such a long journey, you'd need to do 10,000 flights to get to a million people. So 10,000 flights over what period of time? Given that you can only really depart for Mars once every two years, that means you would need like forty or fifty years.

"And then I think for each flight that departs to Mars you want to sort of launch the spacecraft into orbit and then have it be in a parking orbit and refuel its tanks with propellant. Essentially, the spacecraft would use a bunch of its propellant to get to orbit, but then you send up a tanker spacecraft to fill up the propellant tanks of the spacecraft so that it can depart for Mars at high speed and can do so and get there in three months instead of six months and with a large payload. I don't have a detailed plan for

Mars but I know of something at least that would work, which is sort of this all-methane system with a big booster, a spacecraft, and a tanker potentially. I think SpaceX will have developed a booster and spaceship in the 2025 time frame capable of taking large quantities of people and cargo to Mars.

"The thing that's important is to reach an economic threshold around the cost per person for a trip to Mars. If it costs $1 billion per person, there will be no Mars colony. At around $1 million or $500,000 per person, I think it's highly likely that there will be a self-sustaining Martian colony. There will be enough people interested who will sell their stuff on Earth and move. It's not about tourism. It's like people coming to America back in the New World days. You move, get a job there, and make things work. If you solve the transport problem, it's not that hard to make a pressurized transparent greenhouse to live in. But if you can't get there in the first place, it doesn't matter.

"Eventually, you'd need to heat Mars up if you want it to be an Earthlike planet, and I don't have a plan for that. That would take a long time in the best of circumstances. It would probably take, I don't know, somewhere between a century and a millennium. There's zero chance of it being terraformed and Earthlike in my lifetime. Not zero, but 0.001 percent chance, and you would have to take real drastic measures with Mars.'"

---

* For the space buffs, here's Musk talking more about the physics and chemistry of the spaceship: "The final piece of the puzzle for figuring out the Mars architecture is a methane engine. You need to be able to generate the propellant on the surface. Most of the fuel used in rockets today is a form of kerosene, and creating kerosene is quite complex. It's a series of long-chain hydrocarbons. It's much easier to create either methane or hydrogen. The problem with hydrogen is it's a deep cryogen. It's only a liquid very close to absolute zero. And because it's a small molecule you have these issues where hydrogen will seep its way through a metal matrix and embrittle or destroy

Musk spent months pacing around his home in Los Angeles late at night thinking about these plans for Mars and bounc-
ing them off Riley, whom he remarried near the end of 2012.* "I
mean, there aren't that many people you can talk to about this sort
of thing," Musk said. These chats included Musk daydreaming
aloud about becoming the first man to set foot on the Red Planet.

---

metal in weird ways. Hydrogen's density is also very porous, so the
tanks are enormous and it's expensive to create and store hydrogen.
It's not a good choice as a fuel.

"Methane, on the other hand, is much easier to handle. It's liquid at
around the same temperature as liquid oxygen so you can do a rocket
stage with a common bulkhead and not worry about freezing one or
the other solid. Methane is also the lowest-cost fossil fuel on Earth.
And there needs to be a lot of energy to go to Mars.

"And then on Mars, because the atmosphere is carbon dioxide and
there's a lot of water or ice in the soil, the carbon dioxide gets you CO_2,
the water gives you H_2O. With that you create CH_4 and O_2, which
gives you combustion. So it's all sort of nicely worked out.

"And then one of the key questions is can you get to the surface of
Mars and back to Earth on a single stage. The answer is yes, if you
reduce the return payload to approximately one-quarter of the out-
bound payload, which I thought made sense because you are going to
want to transport a lot more to Mars than you'd want to transfer from
Mars to Earth. For the spacecraft, the heat shield, the life support
system, and the legs will have to be very, very light."

* Musk and Riley were divorced for less than year. "I refused to speak
with him for as long it took for the divorce to be finalized," Riley said.
"And then, once it was finalized, we immediately got back together."
As for what caused the breakup, Riley said, "I just wasn't happy. I
thought maybe I had made the wrong decision for my life." And, about
what brought her back to Musk, Riley said, "One reason was the lack
of viable alternatives. I looked around, and there was no one else nice
to be with. Number two is that Elon doesn't have to listen to anyone in
life. No one. He doesn't have to listen to anything that doesn't fit into
his worldview. But he proved he would take shit from me. He said, 'Let
me listen to her and figure these things out.' He proved that he valued
my opinion on things in life and was willing to listen. I thought it was
quite a telling thing for the man—that he made the effort. And then,
I loved him and missed him."

"He definitely wants to be the first man on Mars," Riley said. "I have begged him not to be." Perhaps Musk enjoys teasing his wife or maybe he's playing coy, but he denied this ambition during one of our late-night chats. "I would only be on the first trip to Mars if I was confident that SpaceX would be fine if I die," he said. "I'd like to go, but I don't have to go. The point is not about me visiting Mars but about enabling large numbers of people to go to the planet." Musk may not even go into space. He does not plan to participate in SpaceX's upcoming human test flights. "I don't think that would be wise," he said. "It would be like the head of Boeing being a test pilot for a new plane. It's not the right thing for SpaceX or the future of space exploration. I might be on there if it's been flying for three or four years. Honestly, if I never go to space, that will be okay. The point is to maximize the probable life span of humanity."

It's difficult to gauge just how seriously the average person takes Musk when he talks like this. A few years ago, most people would have lumped him into the category of people who hype up jet packs and robots and whatever else Silicon Valley decided to fixate on for the moment. Then Musk filed away one accomplishment after another, transforming himself from big talker to one of Silicon Valley's most revered doers. Thiel has watched Musk go through this maturation—from the driven but insecure CEO of PayPal to a confident CEO who commands the respect of thousands. "I think there are ways he has dramatically improved over time," said Thiel. Most impressive to Thiel has been Musk's ability to find bright, ambitious people and lure them to his companies. "He has the most talented people in the aerospace industry working for him, and the same case can be made for Tesla, where, if you're a talented mechanical engineer who likes building cars, then you're going to Tesla because it's probably the only company in the U.S. where you can do interesting new things. Both companies were designed with this vision of motivating a critical mass of talented people to work on inspiring things." Thiel thinks Musk's goal of getting humans to Mars should be taken seriously and believes it gives the public hope. Not everyone will identify with the mission but the fact that there's someone out there pushing exploration and our technical abilities to their limits is important. "The goal of sending a man to Mars is so much more inspiring than what other people are trying to do in space," Thiel said. "It's this going-back-to-the-future idea. There's been this long wind-down of the space program, and people have abandoned the optimistic visions of the future that we had in the early 1970s. SpaceX shows there is a way toward bringing back that future. There's great value in what Elon is doing."

The true believers came out in full force in August 2013 when Musk unveiled something called the Hyperloop. Billed as a new mode of transportation, this machine was a large-scale pneumatic tube like the ones used to send mail around offices. Musk proposed linking cities like Los Angeles and San Francisco via an elevated version of this kind of tube that would transport people and cars in pods. Similar ideas had been proposed before, but Musk's creation had some unique elements. He called for the tube to run under low pressure and for the pods to float on a bed of air produced by skis at their base. Each pod would be thrust forward by an electromagnetic pulse, and motors placed throughout the tube would give the pods added boosts as needed. These mechanisms could keep the pods going at 800 mph, allowing someone to travel from Los Angeles to San Francisco in about thirty minutes. The whole thing would, of course, be solar-powered and aimed at linking cities less than a thousand miles apart. "It makes sense for things like L.A. to San Francisco, New York to D.C., New York to Boston," Musk said at the time. "Over one thousand miles, the tube cost starts to become prohibitive, and you don't want tubes every which way. You don't want to live in Tube Land."

Musk had been thinking about the Hyperloop for a number of months, describing it to friends in private. The first time he talked about it to anyone outside of his inner circle was during one of our interviews. Musk told me that the idea originated out of his hatred for California's proposed high-speed rail system. "The sixty-billion-dollar bullet train they're proposing in California would be the slowest bullet train in the world at the highest cost per mile," Musk said. "They're going for records in all the wrong ways." California's high-speed rail is meant to allow people to go from Los Angeles to San Francisco in about two and a half hours upon its completion in—wait for it—2029. It takes about an hour to fly between the cities today and five hours to drive, placing the train right in the zone of mediocrity, which particularly gnawed at Musk. He insisted the Hyperloop would cost about $6 billion to $10 billion, go faster than a plane, and let people drive their cars onto a pod and drive out into a new city.

At the time, it seemed that Musk had dished out the Hyperloop proposal just to make the public and legislators rethink the high-speed train. He didn't actually intend to build the thing. It was more that he wanted to show people that more creative ideas were out there for things that might actually solve problems and push the state forward. With any luck, the high-speed rail would be canceled. Musk said as much to me during a series of e-mails and phone calls leading up to the announcement. "Down the road, I might fund or advise on a Hyperloop project, but right now I can't take my eye off the ball at either SpaceX or Tesla," he wrote.

Musk's tune, however, started to change after he released the paper detailing the Hyperloop. *Bloomberg Businessweek* had the first story on it, and the magazine's Web server began melting down as people stormed the website to read about the invention. Twitter went nuts as well. About an hour after Musk released the information, he held a conference call to talk about the Hyperloop, and somewhere in between our numerous earlier chats and that moment, he'd decided to build the thing, telling reporters that he would consider making at least a prototype to prove that the technology could work. Some people had their fun with all of this. "Billionaire unveils imaginary space train," teased Valleywag. "We love Elon Musk's nutso determination—there was certainly a time when electric cars and private space flight seemed silly, too. But what's sillier is treating this as anything other than a very rich man's wild imagination." Unlike its early Tesla-bashing days, Valleywag was now the minority voice. People seemed mainly to believe Musk could do it. The depth to which people believed it, I think, surprised Musk and forced him to commit to the prototype. In a weird life-imitating-art moment, Musk really had become the closest thing the world had to Tony Stark, and he could not let his adoring public down.

Shortly after the release of the Hyperloop plans, Shervin Pishevar, an investor and friend of Musk's, brought the detailed specifications for the technology with him during a ninety-minute meeting with President Obama at the White House. "The president fell in love with the idea," Pishevar said. The president's staff studied the documents and arranged a one-on-one with Musk and Obama in April 2014. Since then, Pishevar, Kevin Brogan, and others, have formed a company called Hyperloop Technologies Inc. with the hopes of building the first leg of the Hyperloop between Los Angeles and Las Vegas. In theory, people would be able to hop between the two cities in about ten minutes. Nevada senator Harry Reid has been briefed on the idea as well, and efforts are under way to buy the land rights alongside

Interstate 15 that would make the high-speed transport possible.

For employees like Gwynne Shotwell and J. B. Straubel, working with Musk means helping develop these sorts of wonderful technologies in relative obscurity. They're the steady hands that will forever be expected to stay in the shadows. Shotwell has been a consistent presence at SpaceX almost since day one, pushing the company forward and suppressing her ego to ensure that Musk gets all the attention he desires. If you're Shotwell and truly believe in the cause of sending people to Mars, then the mission takes precedence over personal desires. Straubel, likewise, has been the constant at Tesla—a go-between whom other employees could rely on to carry messages to Musk, and the guy who knows everything about the cars. Despite his stature at the company, Straubel was one of several longtime employees who confessed they were nervous to speak with me on the record. Musk likes to be the guy talking on his companies' behalf and comes down hard on even his most loyal executives if they say something deemed to be out of line with Musk's views or with what he wants the public to think. Straubel has dedicated himself to making electric cars and didn't want some dumb reporter wrecking his life's work. "I try really hard to back away and put my ego aside," Straubel said. "Elon is incredibly difficult to work for, but it's mostly because he's so passionate. He can be impatient and say, 'God damn it! This is what we have to do!' and some people will get shell-shocked and catatonic. It seems like people can get afraid of him and paralyzed in a weird way. I try to help everyone to understand what his goals and visions are, and then I have a bunch of my own goals, too, and make sure we're in synch. Then, I try and go back and make sure the company is aligned. Ultimately, Elon is the boss. He has driven this thing with his blood, sweat, and tears. He has risked more than anyone else. I respect the hell out of what he has done. It

communications staff, they're expected to jump in without missing a beat and to execute at the highest level. Some of this staff, operating under this mix of pressure and surprise, only lasted between a few weeks and a few months. A few others have hung on for a couple of years before burning out or being fired.

The granddaddy example of Musk's seemingly callous interoffice style occurred in early 2014 when he fired Mary Beth Brown. To describe her as a loyal executive assistant would be grossly inadequate. Brown often felt like an extension of Musk—the one being who crossed over into all of his worlds. For more than a decade, she gave up her life for Musk, traipsing back and forth between Los Angeles and Silicon Valley every week, while working late into the night and on weekends. Brown went to Musk and asked that she be compensated on par with SpaceX's top executives, since she was handling so much of Musk's scheduling across two companies, doing public relations work and often making business decisions. Musk replied that Brown should take a couple of weeks off, and he would take on her duties and gauge how hard they were. When Brown returned, Musk let her know that he didn't need her anymore, and he asked Shotwell's assistant to begin scheduling his meetings. Brown, still loyal and hurt, didn't want to discuss any of this with me. Musk said that she had become too comfortable speaking on his behalf and that, frankly, she needed a life. Other people grumbled that Brown and Riley clashed and that this was the root cause of Brown's ouster, although whether there is any truth in this I do not know.*

---

* As Musk recalled, "I told her, 'Look, I think you're very valuable. Maybe that compensation is right. You need to take two weeks' vacation, and I'm going to assess whether that's true our not.' Before this came up, I had offered her multiple all-expenses-paid vacations. I really wanted her to take a vacation. When she got back, my conclusion was just that the relationship was not going to work anymore. Twelve years is a good run for any job. She'll do a great job for someone." Accord-

(Brown declined to be interviewed for this book, despite several requests.)

Whatever the case, the optics of the situation were terrible. Tony Stark doesn't fire Pepper Potts. He adores her and takes care of her for life. She's the only person he can really trust—the one who has been there through everything. That Musk was willing to let Brown go and in such an unceremonious fashion struck people inside SpaceX and Tesla as scandalous and as the ultimate confirmation of his cruel stoicism. The tale of Brown's departure became part of the lore around Musk's lack of empathy. It got bundled up into the stories of Musk dressing employees down in legendary fashion with vicious barb after vicious barb. People also linked this type of behavior to Musk's other quirky traits. He's been known to obsess over typos in e-mails to the point that he could not see past the errors and read the actual content of the messages. Even in social settings, Musk might get up from the dinner table without a word of explanation to head outside and look at the stars, simply because he's not willing to suffer fools or small talk. After adding up this behavior, dozens of people expressed to me their conclusion that Musk sits somewhere on the autism spectrum and that he has trouble considering other people's emotions and caring about their well-being.

There's a tendency, especially in Silicon Valley, to label people who are a bit different or quirky as autistic or afflicted with Asperger's syndrome. It's armchair psychology for conditions that can be inherently funky to diagnose or even codify. To slap this label on Musk feels ill-informed and too easy.

Musk acts differently with his closest friends and family than he does with employees, even those who have worked alongside him for a long time. Among his inner circle, Musk is warm, funny,

---

ing to Musk, he offered Brown another position at the company. She declined the offer by never showing up at the office again. Musk gave her twelve months' severance and has not spoken to her since.

and deeply emotional.* He might not engage in the standard chit-chat, asking a friend how his kids are doing, but he would do everything in his considerable power to help that friend if his child were sick or in trouble. He will protect those close to him at all costs and, when deemed necessary, seek to destroy those who have wronged him or his friends.

---

* According to Riley, "Elon is kind of cheeky and funny. He is very loving. He is devoted to his children. He is funny—really, really, really funny. He's quite mercurial. He's genuinely the oddest person I have ever met. He has moments of self-awareness and lucidity, which for me always bring him back around. He'll say something cheeky or funny and have this grin. He's smart in all sorts of areas. He's very well read and has this incredible wit. He loves movies. We went to see the new *Lego Movie* and afterwards he insisted on being referred to as Lord Business. He tries to come home early for family dinners with me and the kids and maybe play some computer games with the boys. They will tell us about their day, and we'll put them to bed. Then we'll chat and watch something together on the laptop like *The Colbert Report*. On the weekends, we're traveling. The kids are good travelers. There were bajillions of nannies before. There was even a nanny manager. Things are a bit more normal now. We try and do stuff just as a family when we can. We have the kids four days a week. I like to say that I am the disciplinarian. I want them to have the sense of an ordinary life, but they live a very odd life. They were just on a trip with Justin Bieber. They go to the rocket factory and are like, 'Oh no, not again.' It's not cool if your dad does it. They're used to it.

"People don't realize that Elon has this incredible naiveté. There are certain times when he is incapable of anything other than pure joy. And then other times pure anger. When he feels something, he feels it so completely and purely. Nothing else can impose on it. There are so few people who can do that. If he sees something funny, he will laugh so loudly. He won't realize we are in a crowded theater and that other people are there. He is like a child. It's sweet and amazing. He says this random stuff like, 'I am a complicated man with very simple but specific needs' or 'No man is an island unless he is large and buoyant.' We make these lists of things we want to do. His latest contributions were to walk on a beach at sunset and whisper sweet nothings in each other's ear and to take more horseback rides. He likes reading, playing video games, and being with friends."

Musk's behavior matches up much more closely with someone who is described by neuropsychologists as profoundly gifted. These are people who in childhood exhibit exceptional intellectual depth and max out IQ tests. It's not uncommon for these children to look out into the world and find flaws—glitches in the system—and construct logical paths in their minds to fix them. For Musk, the call to ensure that mankind is a multiplanetary species partly stems from a life richly influenced by science fiction and technology. Equally it's a moral imperative that dates back to his childhood. In some form, this has forever been his mandate.

Each facet of Musk's life might be an attempt to soothe a type of existential depression that seems to gnaw at his every fiber. He sees man as self-limiting and in peril and wants to fix the situation. The people who suggest bad ideas during meetings or make mistakes at work are getting in the way of all of this and slowing Musk down. He does not dislike them as people. It's more that he feels pained by their mistakes, which have consigned man to peril that much longer. The perceived lack of emotion is a symptom of Musk sometimes feeling like he's the only one who really grasps the urgency of his mission. He's less sensitive and less tolerant than other people because the stakes are so high. Employees need to help solve the problems to the absolute best of their ability or they need to get out of the way.

Musk has been pretty up front about these tendencies. He's implored people to understand that he's not chasing momentary opportunities in the business world. He's trying to solve problems that have been consuming him for decades. During our conversations, Musk went back to this very point over and over again, making sure to emphasize just how long he'd thought about electric cars and space. The same patterns are visible in his actions as well. When Musk announced in 2014 that Tesla would open-source all of its patents, analysts tried to decide whether this was a publicity stunt or if it hid an ulterior motive or a catch. But the decision was a straightforward one for Musk. He wants people to make and buy electric cars. Man's future, as he sees it, depends on this. If open-sourcing Tesla's patents means other companies can build electric cars more easily, then that is good for mankind, and the ideas should be free. The cynic will scoff at this, and understandably so. Musk, however, has been programmed to behave this way and tends to be sincere when explaining his thinking—almost to a fault.

The people who get closest to Musk are the ones who learn to relate to this mode of thinking.[22] They're the ones who can identify with his vision yet challenge him intellectually to complete it. When he asked me during one of our dinners if I thought he was insane, it was a test of sorts. We had talked enough that he knew I was interested in what he was doing. He had started to trust me and open up but wanted to make sure—one final time—that I truly grasped the importance of his quest. Many of his closest friends have passed much grander, more demanding tests. They've invested in his companies. They've defended him against critics. They helped him keep the wolves at bay during 2008. They've proven their loyalty and their commitment to his cause.

People in the technology industry have tended to liken Musk's drive and the scope of his ambition to that of Bill Gates and Steve Jobs. "Elon has that deep appreciation for technology, the no-holds-barred attitude of a visionary, and that determination to go after long-term things that they both had," said Edward Jung, a child prodigy who worked for Jobs and Gates and ended up as Microsoft's chief software architect. "And he has that consumer sensibility of Steve along with the ability to hire good people outside of his own comfort areas that's more like Bill. You almost wish that Bill and Steve had a genetically engineered love child and, who knows, maybe we should genotype Elon to see if that's what happened." Steve Jurvetson, the venture capitalist who has invested in SpaceX, Tesla, and SolarCity, worked for Jobs, and knows Gates well, also described Musk as an upgraded mix of the two. "Like Jobs, Elon does not tolerate C or D players," said Jurvetson. "But I'd say he's nicer than Jobs and a bit more refined than Bill Gates."*

But the more you know about Musk, the harder it becomes to place him among his peers. Jobs is another CEO who ran two, large industry-changing companies—Apple and Pixar. But that's where the practical similarities between the two men end. Jobs dedicated far more of his energy to Apple than Pixar, unlike Musk, who has poured equal energy into both companies, while saving whatever was left over for SolarCity. Jobs was also legendary for his attention to detail. No one, however, would suggest that his reach extended down as far as Musk's into overseeing so much of the companies' day-to-day operations. Musk's approach has its limitations. He's less artful with marketing and media strategy. Musk does not rehearse his presentations or polish speeches. He wings most of the announcements from Tesla and SpaceX. He'll also fire off some major bit of news on a Friday afternoon when it's likely to get lost as reporters head home for the weekend, simply because that's when he finished writing the press release or wanted to move on to something else. Jobs, by contrast, treated every presentation and media moment as precious. Musk simply

---

*   Jurvetson elaborated by saying, "Elon has that engineering prowess of Gates, but he's more interpersonal. You have to be out there on the spectrum with Gates. Elon has more interpersonal charms. He's like Jobs in that neither of them suffer fools. But with Jobs there was more of a hero-shit roller coaster where employees went from in favor to out of favor. I also think Elon has accomplished more."

does not have the luxury to work that way. "I don't have days to practice," he said. "I've got to give impromptu talks, and the results may vary."

As for whether Musk is leading the technology industry to new heights like Gates and Jobs, the professional pundits remain mixed. One camp holds that SolarCity, Tesla, and SpaceX offer little in the way of real hope for an industry that could use some blockbuster innovations. For the other camp, Musk is the real deal and the brightest shining star of what they see as a coming revolution in technology.

The economist Tyler Cowen—who has earned some measure of fame in recent years for his insightful writings about the state of the technology industry and his ideas on where it may go—falls into that first camp. In *The Great Stagnation*, Cowen bemoaned the lack of big technological advances and argued that the American economy has slowed and wages have been depressed as a result. "In a figurative sense, the American economy has enjoyed lots of low-hanging fruit since at least the seventeenth century, whether it be free land, lots of immigrant labor, or powerful new technologies," he wrote. "Yet during the last forty years, that low-hanging fruit started disappearing, and we started pretending it was still there. We have failed to recognize that we are at a technological plateau and the trees are more bare than we would like to think. That's it. That is what has gone wrong."

In his next book, *Average Is Over*, Cowen predicted an unromantic future in which a great divide had occurred between the Haves and the Have Nots. In Cowen's future, huge gains in artificial intelligence will lead to the elimination of many of today's high-employment lines of work. The people who thrive in this environment will be very bright and able to complement the machines and team effectively with them. As for the unemployed masses? Well, many of them will eventually find jobs going to work for the Haves, who will employ teams of nannies, house-keepers, and gardeners. If anything Musk is doing might alter the course of mankind toward a rosier future, Cowen can't find it. Coming up with true breakthrough ideas is much harder today than in the past, according to Cowen, because we've already mined the bulk of the big discoveries. During a lunch in Virginia, Cowen described Musk not as a genius inventor but as an atten-tion seeker, and not a terribly good one at that. "I don't think a lot of people care about getting to Mars," he said. "And it seems like a very expensive way to drive whatever breakthroughs you might get from it. Then, you hear about the Hyperloop. I don't think he has any intention of doing it. You have to wonder if it's not meant just to be publicity for his companies. As for Tesla, it might work. But you're still just pushing the problems back somewhere else. You still have to generate power. It could be that he is challenging convention less than people think."

These sentiments are not far off from those of Vaclav Smil, a professor emeritus at the University of Manitoba. Bill Gates has hailed Smil as an important writer for his tomes on energy, the environment, and manufacturing. One of Smil's latest works is *Made in the USA*, an exploration of America's past manufacturing glories and its subsequent, dismal loss of industry. Anyone who thinks the United States is making a natural, clever shift away from manufacturing and toward higher-paying information-worker jobs will want to read this book and have a gander at the long-term consequences of this change. Smil presents numer-ous examples of the ways in which the manufacturing industry generates major innovations and creates a massive ecosystem of jobs and technical smarts around them. "For example, when some three decades ago the United States stopped making virtu-ally all 'commodity' consumer electronic devices and displays, it also lost its capacity to develop and mass-produce advanced flat screens and batteries, two classes of products that are quintessential for portable computers and cell phones and whose large-scale imports keep adding to the US trade deficit," Smil wrote. A bit later in the book, Smil emphasized that the aerospace industry, in particular, has been a huge boon to the U.S. economy and one of its major exporters. "Maintaining the sector's competitiveness must be a key component of efforts to boost US exports, and the exports will have to be a large part of the sector's sales because the world's largest aerospace market of the next two decades will be in Asia, above all in China and India, and American aircraft and aeroengine makers should benefit from this expansion."

Smil is consumed by the United States' waning ability to compete with China and yet does not perceive Musk or his companies as any sort of counter to this slide. "As, among other things, a historian of technical advances I simply must see Tesla as nothing but an utterly derivative overhyped toy for showoffs," Smil wrote to me. "The last thing a country with 50 million people on food stamps and 85 billion dollars deeper into debt every month needs is anything to do with space, especially space with more joyrides for the super rich. And the loop proposal was nothing but bamboozling people who do not know anything about kindergarten physics with a very old, long publicized Gedankenexperiment in kinetics. . . . There are many inventive Americans, but in that lineup Musk would be trailing far behind."

The comments were blunt and surprising given some of the things Smil celebrated in his recent book. He spent a good deal of time showing the positive impact that Henry Ford's vertical integration had on advancing the car industry and the American economy. He also wrote at length about the rise of "mechatronic machines," or machines that rely on a lot of electronics and software. "By 2010 the electronic controls for a typical sedan required more lines of software code than the instructions needed to operate the latest Boeing jetliner," Smil wrote. "American manufacturing has turned modern cars into remarkable mechatronic machines. The first decade of the twenty-first century also brought innovations ranging from the deployment of new materials (carbon composites in aviation, nanostructures) to wireless electronics."

There's a tendency among critics to dismiss Musk as a frivolous dreamer that stems first and foremost from a misunderstanding of what Musk is actually doing. People like Smil seem to catch an article or television show that hits on Musk's quest to get to Mars and immediately lump him with the space tourism crowd. Musk, though, hardly ever talks about tourism and has, since day one, built up SpaceX to compete at the industrial end of the space business. If Smil thinks Boeing selling planes is crucial to the American economy, then he should be enthused about what SpaceX has managed to accomplish in the commercial launch market. SpaceX builds its products in the United States, has made dramatic advances in aerospace technology, and has made similar advances in materials and manufacturing techniques. It would not take much to argue that SpaceX is America's only hope of competing against China in the next couple of decades. As for mechatronic machines, SpaceX and Tesla have set the example of fusing together electronics, software, and metal that their rivals are now struggling to match. And all of Musk's companies, including SolarCity, have made dramatic use of vertical integration and turned in-house control of components into a real advantage.

To get a sense of how powerful Musk's work may end up being for the American economy, have a think about the dominant mechatronic machine of the past several years: the smartphone. Pre-iPhone, the United States was the laggard in the telecommunications industry. All of the exciting cell phones and mobile services were in Europe and Asia, while American consumers bumbled along with dated equipment. When the iPhone arrived in 2007, it changed everything. Apple's device mimicked many of the functions of a computer and then added new abilities with its apps, sensors, and location awareness. Google charged to market with its Android software and related handsets, and the United States suddenly emerged as the driving force in the mobile industry. Smartphones were revolutionary because of the ways they allowed hardware, software, and services to work in unison. This was a mix that favored the skills of Silicon Valley. The rise of the smartphone led to a massive industrial boom in which Apple became the most valuable company in the country, and billions of its clever devices were spread all over the world.

Tony Fadell, the former Apple executive credited with bringing the iPod and iPhone to market, has characterized the smartphone as representative of a type of super-cycle in which hardware and software have reached a critical point of maturity. Electronics are good and cheap, while software is more reliable and sophisticated. Their interplay is now resulting in science fiction–worthy ideas we were promised long ago becoming a reality. Google has its self-driving cars and has acquired dozens of robotics companies as it looks to merge code and machine. Fadell's company Nest has its intelligent thermostats and smoke alarms. General Electric has jet engines packed full of sensors taught to proactively report possible anomalies to its human mechanics. And a host of start-ups have begun infusing medical devices with powerful software to help people monitor and analyze their bodies and diagnose conditions. Tiny satellites are being put into orbit twenty at a time, and instead of being given a fixed task for their entire lifetimes, like their predecessors, they're being reprogrammed on the fly for a wide variety of business and scientific tasks. Zee Aero, a start-up in Mountain View, has a couple of former SpaceX staffers on hand and is working on a secretive new type of transport. A flying car at last? Perhaps.

For Fadell, Musk's work sits at the highest end of this trend. "He could have just made an electric car," Fadell said. "But he did things like use motors to actuate the door handles. He's bringing the consumer electronics and the software together, and the other car companies are trying to figure out a way to get there. Whether it's Tesla or SpaceX taking Ethernet cables and running them inside of rocket ships, you are talking about combining the old-world science of manufacturing with low-cost, consumer-grade technology. You put these things together, and they morph into something we have never seen before. All of a sudden there is a wholesale change," he said. "It's a step function."

To the extent that Silicon Valley has searched for an inheritor to Steve Jobs's role as the dominant, guiding force of the technology industry, Musk has emerged as the most likely candidate. He's certainly the "it" guy of the moment. Start-up founders, proven executives, and legends hold him up as the person they most admire. The more mainstream Tesla can become, the more Musk's reputation will rise. A hot-selling Model 3 would certify Musk as that rare being able to rethink an industry, read consumers, and execute. From there, his more fanciful ideas start to seem inevitable. "Elon is one of the few people that I feel is more accomplished than I am," said Craig Venter, the man who decoded the human genome and went on to create synthetic life-forms. At some point he hopes to work with Musk on a type of DNA printer that could be sent to Mars. It would, in theory, allow humans to create medicines, food, and helpful microbes for early settlers of the planet. "I think biological teleportation is what is going to truly enable the colonization of space," he said. "Elon and I have been talking about how this might play out."

One of Musk's most ardent admirers is also one of his best friends: Larry Page, the cofounder and CEO of Google. Page has ended up on Musk's house-surfing schedule. "He's kind of homeless, which I think is sort of funny," Page said. "He'll e-mail and say, 'I don't know where to stay tonight. Can I come over?' I haven't given him a key or anything yet."

Google has invested more than just about any other technology company into Musk's sort of moon-shot projects: self-driving cars, robots, and even a cash prize to get a machine onto the moon cheaply. The company, however, operates under a set of constraints and expectations that come with employing tens of thousands of people and being analyzed constantly by investors. It's with this in mind that Page sometimes feels a bit envious of Musk, who has managed to make radical ideas the basis of his companies. "If you think about Silicon Valley or corporate leaders in general, they're not usually lacking in money," Page said. "If you have all this money, which presumably you're going to give away and couldn't even spend it all if you wanted to, why then are you devoting your time to a company that's not really doing anything good? That's why I find Elon to be an inspiring example. He said, 'Well, what should I really do in this world? Solve cars, global warming, and make humans multiplanetary.' I mean those are pretty compelling goals, and now he has businesses to do that."

"This becomes a competitive advantage for him, too. Why would you want to work for a defense contractor when you can work for a guy who wants to go to Mars and he's going to move heaven and earth to make it happen? You can frame a problem in a way that's really good for the business."

At one point, a quotation from Page made the rounds, saying that he wanted to leave all of his money to Musk. Page felt he was misquoted but stood by the sentiment. "I'm not leaving my money to him at the moment," Page said. "But Elon makes a pretty compelling case for having a multiplanetary society just because, you know, otherwise we might all die, which seems like it would be sad for all sorts of different reasons. I think it's a very doable project, and it's a relatively modest resource that we need to set up a permanent human settlement on Mars. I was just trying to make the point that that's a really powerful idea."

As Page puts it, "Good ideas are always crazy until they're not." It's a principle he's tried to apply at Google. When Page and Sergey Brin began wondering aloud about developing ways to search the text inside of books, all of the experts they consulted said it would be impossible to digitize every book. The Google cofounders decided to run the numbers and see if it was actually physically possible to scan the books in a reasonable amount of time. They concluded it was, and Google has since scanned millions of books. "I've learned that your intuition about things you don't know that much about isn't very good," Page said. "The way Elon talks about this is that you always need to start with the first principles of a problem. What are the physics of it? How much time will it take? How much will it cost? How much cheaper can I make it? There's this level of engineering and physics that you need to make judgments about what's possible and interesting. Elon is unusual in that he knows that, and he also knows business and organization and leadership and governmental issues."

Some of the conversations between Musk and Page take place at a secret apartment Google owns in downtown Palo Alto. It's inside of one of the taller buildings in the area and offers views of the mountains surrounding the Stanford University campus. Page and Brin will take private meetings at the apartment and have their own chef on call to prepare food for guests. When Musk is present, the chats tend toward the absurd and fantastic. "I was there once, and Elon was talking about building an electric jet plane that can take off and land vertically," said George Zachary, the venture capitalist and friend of Musk's. "Larry said the plane should be able to land on ski slopes, and Sergey said it needed to be able to dock at a port in Manhattan. Then they started talking about building a commuter plane that was always circling the Earth, and you'd hop up to it and get places incredibly fast. I thought everyone was kidding, but at the end I asked Elon, 'Are you really going to do that?' And he said, 'Yes.'"

"It's kind of our recreation, I guess," said Page.[23] "It's fun for the three of us to talk about kind of crazy things, and we find stuff that eventually turns out to be real. We go through hundreds or thousands of possible things before arriving at the ones that are most promising."

Page talked about Musk at times as if he were a one-of-a-kind, a force of nature able to accomplish things in the business world that others would never even try. "We think of SpaceX and Tesla as being these tremendously risky things, but I think Elon was going to make them work no matter what. He's willing to suffer some personal cost, and I think that makes his odds actually pretty good. If you knew him personally, you would look back to when he started the companies and say his odds of success would be more than ninety percent. I mean we just have a single proof point now that you can be really passionate about something that other people think is crazy and you can really succeed. And you look at it with Elon and you say, 'Well, maybe it's not luck. He's done it twice. It can't be luck totally.' I think that means it should be repeatable in some sense. At least it's repeatable by him. Maybe we should get him to do more things."

Page holds Musk up as a model he wishes others would emulate—a figure that should be replicated during a time in which the businessmen and politicians have fixated on short-term, inconsequential goals. "I don't think we're doing a good job as a society deciding what things are really important to do," Page said. "I think like we're just not educating people in this kind of general way. You should have a pretty broad engineering and scientific background. You should have some leadership training and a bit of MBA training or knowledge of how to run things, organize stuff, and raise money. I don't think most people are doing that, and it's a big problem. Engineers are usually trained in a very fixed area. When you're able to think about all of these disciplines together, you kind of think differently and can dream of much crazier things and how they might work. I think that's really an important thing for the world. That's how we make progress."

The pressure of feeling the need to fix the world takes its toll on Musk's body. There are times when you run into Musk and he looks utterly exhausted. He does not have bags under his eyes but rather deep, shadowy valleys. During the worst of times, following weeks of sleep deprivation, his eyes seem to have sunk back into his skull. Musk's weight moves up and down with the stress, and he's usually heavier when really overworked. It's funny in a way that Musk spends so much time talking about man's survival but isn't willing to address the consequences of what his lifestyle does to his body. "Elon came to the conclusion early in his career that life is short," Straubel said. "If you really embrace this, it leaves you with the obvious conclusion that you should be working as hard as you can."

Suffering, though, has always been Musk's thing. The kids at school tortured him. His father played brutal mind games. Musk then abused himself by working inhumane hours and forever pushing his businesses to the edge. The idea of work-life balance seems meaningless in this context. For Musk, it's just life, and his wife and kids try to fit into the show where they can. "I'm a pretty good dad," Musk said. "I have the kids for slightly more than half the week and spend a fair bit of time with them. I also take them with me when I go out of town. Recently, we went to the Monaco

Grand Prix and were hanging out with the prince and princess of Monaco. It all seemed quite normal to the kids, and they were blasé about it. They are growing up having a set of experiences that are extremely unusual, but you don't realize experiences are unusual until you are much older. They're just your experiences. They have good manners at meals."

It bothers Musk a bit that his kids won't suffer like he did. He feels that the suffering helped to make him who he is and gave him extra reserves of strength and will. "They might have a little adversity at school, but these days schools are so protective," he said. "If you call someone a name, you get sent home. When I was going to school, if they punched you and there was no blood, it was like, 'Whatever. Shake it off.' Even if there was a little blood, but not a lot, it was fine. What do I do? Create artificial adversity? How do you do that? The biggest battle I have is restricting their video game time because they want to play all the time. The rule is they have to read more than they play video games. They also can't play completely stupid video games. There's one game they downloaded recently called Cookies or something. You literally tap a fucking cookie. It's like a Psych 101 experiment. I made them delete the cookie game. They had to play Flappy Golf instead, which is like Flappy Bird, but at least there is some physics involved."

Musk has talked about having more kids, and it's on this subject that he delivers some controversial philosophizing vis-à-vis the creator of *Beavis and Butt-head*. "There's this point that Mike Judge makes in *Idiocracy*, which is like smart people, you know, should at least sustain their numbers," Musk said. "Like, if it's a negative Darwinian vector, then obviously that's not a good thing. It should be at least neutral. But if each successive generation of smart people has fewer kids, that's probably bad, too. I mean, Europe, Japan, Russia, China are all headed for demographic implosion. And the fact of the matter is that basically the wealthier—basically wealth, education, and being secular are all indicative of low birth rate. They all correlate with low birth rate. I'm not saying like only smart people should have kids. I'm just saying that smart people should have kids as well. They should at least maintain—at least be a replacement rate. And the fact of the matter is that I notice that a lot of really smart women have zero or one kid. You're like, 'Wow, that's probably not good.'"

The next decade of Musk Co. should be quite something. Musk has given himself a chance to become one of the greatest businessmen and innovators of all time. By 2025 Tesla could very well have a lineup of five or six cars and be the dominant force in a booming electric car market. Playing off its current growth rate, SolarCity will have had time to emerge as a massive utility company and the leader in a solar market that had finally lived up to its promise. SpaceX? Well, it's perhaps the most intriguing. According to Musk's calculations, SpaceX should be conducting weekly flights to space, carrying humans and cargo, and have put most of its competitors out of business. Its rockets should be capable of doing a couple of laps around the moon and then landing with pinpoint accuracy back at the spaceport in Texas. And the preparation for the first few dozen trips to Mars should be well under way.

If all of this were taking place, Musk, then in his mid-fifties, likely would be the richest man in the world and among its most powerful. He would be the majority shareholder in three public companies, and history would be preparing to smile broadly on what he had accomplished. During a time in which countries and other businesses were paralyzed by indecision and inaction, Musk would have mounted the most viable charge against global warming, while also providing people with an escape plan—just in case. He would have brought a substantial amount of crucial manufacturing back to the United States while also providing an example for other entrepreneurs hoping to harness a new age of wonderful machines. As Thiel said, Musk may well have gone so far as to give people hope and to have renewed their faith in what technology can do for mankind.

This future, of course, remains precarious. Huge technological issues confront all three of Musk's companies. He's bet on the inventiveness of man and the ability of solar, battery, and aerospace technology to follow predicted price and performance curves. Even if these bets hit as he hopes, Tesla could face a weird, unexpected recall. SpaceX could have a rocket carrying humans blow up—an incident that could very well end the company on the spot. Dramatic risks accompany just about everything Musk does.

By the time our last dinner had come around, I had decided that this propensity for risk had little to do with Musk being insane, as he had wondered aloud several months earlier. No, Musk just seems to possess a level of conviction that is so intense and exceptional as to be off-putting to some. As we shared some chips and guacamole and cocktails, I asked Musk directly just how much he was willing to put on the line. His response? Everything that other people hold dear. "I would like to die on Mars," he said. "Just not on impact. Ideally I'd like to go for a visit, come back for a while, and then go there when I'm like seventy or something and then just stay there. If things turn out well, that would be the case. If my wife and I have a bunch of kids, she would probably stay with them on Earth."

EPILOGUE

E LON MUSK IS A BODY THAT REMAINS VERY MUCH IN MOTION.
By the time this book reaches your hands, it's quite possible that Musk and SpaceX will have managed to land a rocket on a barge at sea or back on a launchpad in Florida. Tesla Motors may have unveiled some of the special features of the Model X. Musk could have formally declared war on the artificial intelligence machines coming to life inside of Google's data centers. Who knows?

What's clear is that Musk's desire to take on more keeps growing. Just as I was putting the finishing touches on this book, Musk unfurled a number of major initiatives. The most dramatic of which is a plan to surround the Earth with thousands of small communications satellites. Musk wants, in effect, to build a space-based Internet in which the satellites would be close enough to the planet to beam down bandwidth at high speeds. Such a system would be useful for a couple of reasons: In areas too poor or too remote to have fiber-optic connections, it would provide people with high-speed Internet for the first time. It could also function as an efficient backhaul network for businesses and consumers.

Musk, of course, also sees this space Internet as key to his long-term ambitions around Mars. "It will be important for Mars to have a global communications network," he said. "I think this needs to be done, and I don't see anyone else doing it." SpaceX will build these satellites at a new factory and will also look to sell more satellites to commercial customers as it perfects the technology. To fund part of this unbelievably ambitious project, SpaceX secured $1 billion from Google and Fidelity. In a rare moment of restraint, Musk declined to provide an exact delivery date for his space Internet, which he forecasts will cost more than $10 billion to build. "People should not expect this to be active sooner than five years," he said. "But we see it as a long-term revenue source for SpaceX to be able to fund a city on Mars."

Meanwhile, SolarCity has purchased a new research and development facility near the Tesla factory in Silicon Valley that's intended to aid its manufacturing work. The building it acquired was the old Solyndra manufacturing plant—another symbol of Musk's ability to thrive in the green technology industry that has destroyed so many other entrepreneurs. And Tesla continues to build its Gigafactory in Nevada at pace, while its network of charging stations has saved upward of four million gallons of gas. During a quarterly earnings announcement, J. B. Straubel promised that Tesla would start producing battery systems for home use in 2015 that would let people hop off the grid for periods of time. Musk then one-upped Straubel, bragging that he thinks Tesla could eventually be more valuable than Apple and could challenge it in the race to be the first $1 trillion company. A handful of groups have also set to work building prototype Hyperloop systems in and around California. Oh, and Musk starred in an episode of *The Simpsons* titled "The Musk Who Fell to Earth," in which Homer became his inventive muse.

The heady expansion plans and triumphant rhetoric from Musk were still not quite enough to hide all of Musk Co.'s flaws. Early 2015 marked the vociferous return of Musk's detractors on Wall Street. Tesla's sales in China were lackluster by any measure, and some analysts renewed their doubts about how much long-term demand there would be for the Model S. Tesla's shares slumped and, for the first time in a while, Musk sounded flustered trying to defend the company's position.

The personal costs of Musk's lifestyle were more severe. Musk announced that, once again, he would be divorcing Talulah Riley. According to Musk, Riley wanted a simpler, smaller life in England and had come to despise Los Angeles. "Tried to talk her out of it, but she insisted," Musk told me. "It is possible that she will change her mind at some point, but not anytime soon."

After finishing my reporting and writing for this book, I had a chance to speak with some of Musk's confidantes and employees in a more relaxed manner and bounce various ideas off of them. I'm more convinced than ever that Musk is, and has always been, a man on a quest, and that his brand of quest is far more fantastic and consuming than anything most of us will ever experience. It seems that he's become almost addicted to expanding his ambitions and can't quite stop himself from announcing things like the Hyperloop and the space Internet. I'm also more convinced than ever that Musk is a deeply emotional person who suffers and rejoices in an epic fashion. This side of him is likely obscured by the fact that he feels most deeply about his own humanity-altering quest and so has trouble recognizing the strong emotions of those around him. This tends to make Musk come off as aloof and hard. I would argue, however, that his brand of empathy is unique. He seems to feel for the human species as a whole without always wanting to consider the wants and needs of individuals. And it may well be the case that this is exactly the type of person it takes to make a freaking space Internet real.

THE TECHNOLOGY INDUSTRY LOVES MESSY FOUNDING TALES. A BIT OF backstabbing? A hearty helping of deceit? Perfect. And yet, the press has never really dug into the alleged intrigue surrounding Musk's formation of Zip2, nor have reporters examined the very serious allegations of inconsistencies in Musk's academic record.

In April 2007, a physicist named John O'Reilly filed a lawsuit alleging that Musk had stolen the idea for Zip2. According to the lawsuit, filed with the Superior Court of California in Santa Clara, O'Reilly first met Musk in October 1995. O'Reilly had started a company called Internet Merchant Channel, or IMC, which planned to let businesses create primitive, information-packed online ads. A restaurant, for example, could build an ad that would display its menu and perhaps even turn-by-turn directions to its location. O'Reilly's ideas were mostly theoretical, but Zip2 did end up providing a very similar service. O'Reilly alleged that Musk had first heard about this type of technology while trying to get a job working as a salesman for IMC. He and Musk met on at least three occasions, according to the lawsuit, to talk about the job. O'Reilly then went on an overseas trip and struggled to get back in touch with Musk upon his return.

O'Reilly declined to discuss his case against Musk with me.

But in the lawsuit, he claimed to have learned about Zip2 through happenstance many years after meeting Musk. While reading a book in 2005 about the Internet economy, O'Reilly stumbled upon a passage that mentioned Musk's founding of Zip2 and its 1999 sale to Compaq Computer for $307 million in cash. The physicist was blown away as he realized that Zip2 sounded a lot like IMC, which had never amounted to much of a business. O'Reilly's mind raced back to his encounters with Musk. He began to suspect that Musk had avoided him on purpose and that instead of becoming an IMC salesman, Musk had run off to pursue the same concept on his own. O'Reilly wanted to be compensated for coming up with the original business idea. He spent about two years making his case against Musk. The case file at the court runs hundreds of pages. O'Reilly has affidavits from people that back up parts of his version of events. A judge, however, found that O'Reilly lacked the necessary legal standing to bring this case against Musk due to issues around how his businesses had been dissolved. The judge ordered O'Reilly to shell out $125,000 for Musk's legal fees in 2010. All these years later, Musk still hasn't made O'Reilly pay.

While playing detective, O'Reilly unearthed some information about Musk's past that's arguably more interesting than the allegations in the lawsuit. He found that the University of Pennsylvania granted Musk's degrees in 1997—two years later than what Musk has cited. I called Penn's registrar and verified these findings. Copies of Musk's records show that he received a dual degree in economics and physics in May 1997. O'Reilly also subpoenaed the registrar's office at Stanford to verify Musk's admittance in 1995 for his doctorate work in physics. "Based on the information you provided, we are unable to locate a record in our office for Elon Musk," wrote the director of graduate admissions. When asked during the case to produce a document verifying Musk's enrollment at Stanford, Musk's attorney declined and called the request "unduly burdensome." I contacted a number of Stanford physics professors who taught in 1995, and they either failed to respond or didn't remember Musk. Doug Osheroff, a Nobel Prize winner and department chair at the time, said, "I don't think I knew Elon, and am pretty sure that he was not in the Physics Department."

In the years that have followed, Musk's enemies have been quick to bring up the ambiguities around his admission to Stanford. When Martin Eberhard sued Musk, his attorney introduced O'Reilly's research into the case. And during the course of my interviews, a number of Musk's detractors from the Zip2, PayPal, and early Tesla days said flat out that they think Musk fibbed about getting into Stanford in a bid to boost his credentials as a fledgling entrepreneur and then had to stick with the story after Zip2 took off.

At first, I, too, felt like there were a lot of oddities surrounding Musk's academic record, particularly the Stanford days. But, as I dug in, there were solid explanations for all of the inconsistencies and plenty of evidence to undermine the cases of Musk's detractors.

During the course of my reporting, for example, I found evidence that contradicted O'Reilly's timeline of events. Peter Nicholson, the banker whom Musk had worked for in Canada, took a stroll with Musk along the boardwalk in Toronto before Musk left for Stanford and chatted about the incarnations of something like Zip2. Musk had already started writing some of the early software to support the idea he'd outlined to Kimbal. "He was agonizing whether to do a PhD at Stanford or take this piece of software he'd made in his spare time and make a business out of it," Nicholson said. "He called the thing the Virtual City Navigator. I told him there was this crazy Internet thing going on, and that people will pay big money for damn near anything. This software was a golden opportunity. He could do a PhD anytime." Kimbal and other members of Musk's family have similar memories.

Musk, speaking at length for the first time on the subject, denied everything alleged by O'Reilly and does not even recall meeting the man. "O'Reilly is like a failed physicist who became a serial litigate. And I told the guy, 'Look, I'm not going to settle an unjust case. So it's just like don't even try.' But he still kept at it. His case was tossed out twice on demur, which means that basically even if all the facts in his case were true, he would still lose.

"He'd tried his best to like torture me through my friends and personally [by filing the lawsuit]. And then we've got summary judgment. He lost the summary judgment. He appealed summary judgment, then several months later lost the appeal and I was like, 'Okay, fuck it. Let's file for fees.' And we were awarded fees from when he appealed. And that's when we sent the sheriff after him and he claimed that he had no money basically. Whether he did or didn't I don't know. He certainly claimed he had no money. So we were like either we've got to like impound his car or tap his wife's income. Those didn't seem like great choices. So, we decided that he doesn't have to pay back the money he owes me, so long as he doesn't sue anyone else on frivolous grounds. And, in fact, late last year or early this year [2014], he tried to do just that thing. But, whoever he sued was aware of the nature of my judgment and contacted the lawyer I used, who then told O'Reilly, 'Look, you need to drop the case against these guys or everyone's going to ask for the money. It's kind of pointless to sue them on frivolous grounds because you're going to have fork over the winnings to Elon.' It's like go do something productive with your life."

As for his academic records, Musk produced a document

for me dated June 22, 2009, that came from Judith Haccou, the director of graduate admissions in the office of the registrar at Stanford University. It read, "As per special request from my colleagues in the School of Engineering, I have searched Stanford's admission data base and acknowledge that you applied and were admitted to the graduate program in Material Science Engineering in 1995. Since you did not enroll, Stanford is not able to issue you an official certification document."

Musk also had an explanation for the weird timing on his degrees from Penn. "I had a History and an English credit that I agreed with Penn that I would do at Stanford," he said. "Then I put Stanford on deferment. Later, Penn's requirements changed so that you don't need the English and History credit. So then they awarded me the degree in '97 when it was clear I was not going to go to grad school, and their requirement was no longer there.

"I finished everything that was needed for a Wharton degree in '94. They'd actually mailed me a Wharton degree. I decided to spend another year and finished the physics degree, but then there was that History and English credit thing. I was only reminded about the History and English thing when I tried to get an H-1B visa and called the school to get a copy of my graduation certificate, and they said I hadn't graduated. Then they looked into the new requirements, and said it was fine."

WHILE MUSK HAS REFLECTED PUBLICLY ABOUT HIS TIME AT PAYPAL AND the coup, he went into far greater detail than ever before during one of our longer interviews. Years had passed since the tumultuous days surrounding his ouster, and Musk had been able to meditate more on what went right, what went wrong, and what might have been. He started by discussing his decision to go out of the country, mixing business with a delayed honeymoon, and ended with an explanation of how the finance industry still hasn't solved the problems X.com wanted to tackle.

"The problem with me going away was that I was not there to reassure the board on a few things. Like, the brand change, I think it would have been the right move, but it didn't need to happen right then. At the time it was this weird almost hybrid brand with X.com and PayPal. I think X was the right long-term brand for something that wants to be the central place where all transactions happen. That's the X. It's like the X is the transaction. PayPal doesn't make sense in that context, when we're talking about something more than a personal payment system. I think X was the more sensible approach but timing-wise it didn't need to happen then. That should have probably waited longer.

"As for the technology change, that's not really well understood. On the face of it, it doesn't sound like it makes much sense for us to be writing our front-end code in Microsoft C++ instead of Linux. But the reason is that the programming tools for Microsoft and a PC are actually extremely powerful. They're developed for the gaming industry. I mean, this is going to sound like heresy in a sort of Silicon Valley context, but you can program faster, you can get functionality faster in the PC C++ world. All of the games for the Xbox are written in Microsoft C++. The same goes for games on the PC. They're incredibly sophisticated, hard things to do, and these great tools have been developed thanks to the gaming industry. There were more smart programmers in the gaming industry than anywhere else. I'm not sure the general public understands this. It was also 2000, and there were not the huge software libraries for Linux that you would find today. Microsoft had huge support libraries. So you could get a DLL that could do anything, but you couldn't get—you couldn't get Linux libraries that could do anything.

"Two of the guys that left PayPal went off to Blizzard and helped created World of Warcraft. When you look at the complexity of something like that living on PCs and Microsoft C++, it's pretty incredible. It blows away any website.

"In retrospect, I should have delayed the brand transition, and I should have spent a lot more time with Max getting him comfortable on the technology. I mean, it was a little difficult because like the Linux system Max had created was called Max Code. So Max has had quite a strong affinity for Max Code. This was a bunch of libraries that Max and his friends had done. But it just made it quite hard to develop new features. And if you look at PayPal today, I mean, part of the reason they haven't developed any new features is because it's quite difficult to maintain the old system.

"Ultimately, I didn't disagree with the board's decision in the PayPal case, in the sense that with the information that the board had I would have made maybe the same decision. I probably would have, whereas in the case of Zip2 I would not have. I thought they just simply made a terrible decision based on information they had. I don't think the X.com board made a terrible decision based on the information they had. But it did make me want to be careful about who invested in my companies in the future.

"I've thought about trying to get PayPal back. I've just been too strung out with other things. Almost no one understands how PayPal actually worked or why it took off when other payment systems before and after it didn't. Most of the people at PayPal don't understand this. The reason it worked was because the cost of transactions in PayPal was lower than any other system. And the reason the cost of transactions was lower is because we were able to do an increasing percentage of our transactions as ACH, or automated clearinghouse, electronic transactions, and most importantly, internal transactions. Internal transactions were essentially fraud-free and cost us nothing. An ACH transaction costs, I don't know, like twenty cents or something. But it was slow, so that was the bad thing. It's dependent on the bank's batch processing time. And then the credit card transaction was fast, but expensive in terms of the credit card processing fees and very prone to fraud. That's the problem Square is having now.

"Square is doing the wrong version of PayPal. The critical thing is to achieve internal transactions. This is vital because they are instant, fraud-free, and fee-free. If you're a seller and have various options, and PayPal has the lowest fees and is the most secure, it's obviously the right thing to use.

"When you look at like any given business, like say a business is making 10 percent profitability. They're making 10 percent profit when they may net out all of their costs. You know, revenue minus expenses in a year, they're 10 percent. If using PayPal means you pay 2 percent for your transactions and using some other systems means you pay 4 percent, that means using PayPal gives you a 20 percent increase in your profitability. You'd have to be brain dead not to do that. Right?

"So because about half of PayPal's transactions in the summer of 2001 were internal or ACH transactions, then our fundamental costs of transactions were half because we'd have half credit cards, we'd have that and then the other half would be free. The question then is how do you give people a reason to keep money in the system.

"That's why we created a PayPal debit card. It's a little counterintuitive, but the easier you make it for people to get money out of PayPal, the less they'll want to do it. But if the only way for them to spend money or access it in any way is to move it to a traditional bank, that's what they'll do instantly. The other thing was the PayPal money market fund. We did that because if you consider the reasons that people might move the money out, well, they'll move it to either conduct transactions in the physical world or because they're getting a higher interest rate. So I instituted the highest-return money market fund in the country. Basically, the money market fund was at cost. We didn't intend to make any money on it, in order to encourage people to keep their money in the system. And then we also had like the ability to pay regular bills like your electricity bill and that kind of thing on PayPal.

"There were a bunch of things that should have been done like checks. Because even though people don't use a lot of checks they still use some checks. So if you force people to say, 'Okay, we're not going to let you use checks ever,' they're like, 'Okay, I guess I have to have a bank account.' Just give them a few checks, for God's sake.

"I mean, it's so ridiculous that PayPal today is worse than PayPal circa end of 2001. That's insane.

"None of these start-ups understand the objective. The objective should be—what delivers fundamental value. I think it's important to look at things from a standpoint of what is actually the best thing for the economy. If people can conduct their transactions quickly and securely that's better for them. If it's simpler to conduct their financial life it's better for them. So, if all your financial affairs are seamlessly integrated in one place it's very easy to do transactions and the fees associated with transactions are low. These are all good things. Why aren't they doing this? It's mad."

APPENDIX 3

From: Elon Musk
Date: June 7, 2013, 12:43:06 AM PDT
To: All <All@spacex.com>
Subject: Going Public

Per my recent comments, I am increasingly concerned about SpaceX going public before the Mars transport system is in place. Creating the technology needed to establish life on Mars is and always has been the fundamental goal of SpaceX. If being a public company diminishes that likelihood, then we should not do so until Mars is secure. This is something that I am open to reconsidering, but, given my experiences with Tesla and SolarCity, I am hesitant to foist being public on SpaceX, especially given the long term nature of our mission.

Some at SpaceX who have not been through a public company experience may think that being public is desirable. This is not so. Public company stocks, particularly if big step changes in technology are involved, go through extreme volatility, both for reasons of internal execution and for reasons that have nothing to do with anything except the economy. This causes people to be distracted by the manic-depressive nature of the stock instead of creating great products.

It is important to emphasize that Tesla and SolarCity are public because they didn't have any choice. Their private capital structure was becoming unwieldy and they needed to raise a lot of equity capital. SolarCity also needed to raise a huge amount of debt at the lowest possible interest rate to fund solar leases. The banks who provide that debt wanted SolarCity to have the additional and painful scrutiny that comes with being public. Those rules, referred to as Sarbanes-Oxley, essentially result in a tax being levied on company execution by requiring detailed reporting right down to how your meal is expensed during travel and you can be penalized even for minor mistakes.

YES, BUT I COULD MAKE MORE MONEY IF WE WERE PUBLIC

For those who are under the impression that they are so clever that they can outsmart public market investors and would sell SpaceX stock at the "right time," let me relieve you of any such notion. If you really are better than most hedge fund managers, then there is no need to worry about the value of your SpaceX stock, as you can just invest in other public company stocks and make billions of dollars in the market.

If you think: "Ah, but I know what's really going on at SpaceX and that will give me an edge," you are also wrong. Selling public company stock with insider knowledge is illegal. As a result, selling public stock is restricted to narrow time windows a few times per year. Even then, you can be prosecuted for insider trading. At Tesla, we had both an employee and an investor go through a grand jury investigation for selling stock over a year ago, despite them doing everything right in both the letter and spirit of the law. Not fun.

Another thing that happens to public companies is that you become a target of the trial lawyers who create a class action lawsuit by getting someone to buy a few hundred shares and then suing the company on behalf of all investors for any drop in the stock price. Tesla is going through that right now even though the stock price is relatively high, because the drop in question occurred last year.

It is also not correct to think that because Tesla and SolarCity share prices are on the lofty side right now, that SpaceX would be too. Public companies are judged on quarterly performance. Just because some companies are doing well, doesn't mean that all would. Both of those companies (Tesla in particular) had great first quarter results. SpaceX did not. In fact, financially speaking, we had an awful first quarter. If we were public, the short sellers would be hitting us over the head with a large stick.

We would also get beaten up every time there was an anomaly on the rocket or spacecraft, as occurred on flight 4 with the engine failure and flight 5 with the Dragon prevalves. Delaying launch of V1.1, which is now over a year behind schedule, would result in particularly severe punishment, as that is our primary revenue driver. Even something as minor as pushing a launch back a few weeks from one quarter to the next gets you a spanking. Tesla vehicle production in Q4 last year was literally only three weeks behind and yet the market response was brutal.

BEST OF BOTH WORLDS

My goal at SpaceX is to give you the best aspects of a public and private company. When we do a financing round, the stock price is keyed off of approximately what we would be worth if publicly traded, excluding irrational exuberance or depression, but without the pressure and distraction of being under a hot public spotlight. Rather than have the stock be up during one liquidity window and down during another, the goal is a steady upward trend and never to let the share price go below the last round. The end result for you (or an investor in SpaceX) financially will be the same as if we were public and you sold a steady amount of stock every year.

In case you are wondering about a specific number, I can say that I'm confident that our long term stock price will be over $100 if we execute well on Falcon 9 and Dragon. For this to be the case, we must have a steady and rapid cadence of launch that is far better than what we have achieved in the past. We have more work ahead of us than you probably realize. Let me give you a sense of where things stand financially: SpaceX expenses this year will be roug[h]ly $800 to $900 million (which blows my mind btw). Since we get revenue of $60M for every F9 flight or double that for a FH or F9-Dragon flight, we must have about twelve flights per year where four of those flights are either Dragon or Heavy merely in order to achieve 10% profitability!

For the next few years, we have NASA commercial crew funding that helps supplement those numbers, but, after that, we are on our own. That is not much time to finish F9, FH, Dragon V2 and achieve an average launch

rate of at least one per month. And bear in mind that is
an average, so if we take an extra three weeks to launch a
rocket for any reason (could even be due to the satellite), we
have only one week to do the follow-on flight.

MY RECOMMENDATION

Below is my advice about regarding selling SpaceX stock or
options. No complicated analysis is required, as the rules of
thumb are pretty simple.

If you believe that SpaceX will execute better than the
average public company, then our stock price will continue
to appreciate at a rate greater than that of the stock market,
which would be the next highest return place to invest
money over the long term. Therefore, you should sell
only the amount that you need to improve your standard
of living in the short to medium term. I do actually
recommend selling some amount of stock, even if you are
certain it will appreciate, as life is short and a bit more
cash can increase fun and reduce stress at home (so long as
you don't ratchet up your ongoing personal expenditures
proportionately).

To maximize your post tax return, you are probably
best off exercising your options to convert them to stock
(if you can afford to do this) and then holding the stock for
a year before selling it at our roughly biannual liquidity
events. This allows you to pay the capital gains tax rate,
instead of the income tax rate.

On a final note, we are planning to do a liquidity event
as soon as Falcon 9 qualification is complete in one to two
months. I don't know exactly what the share price will be

yet, but, based on initial conversations with investors, I
would estimate probably between $30 and $35. This places
the value of SpaceX at $4 to $5 billion, which is about
what it would be if we were public right now and, frankly,
an excellent number considering that the new F9, FH and
Dragon V2 have yet to launch.

Elon

ACKNOWLEDGMENTS

F ROM A PROCESS PERSPECTIVE, THIS WILL ALWAYS BE TWO BOOKS instead of one in my mind. There's the time Before Elon, and the time After Elon.

The first eighteen months or so of reporting were filled with tension, sorrow, and joy. As mentioned in the main text, Musk initially opted against helping me with the project. This left me going from interview subject to interview subject, giving a huge windup each time to try to talk an ex-Tesla employee or an old schoolmate into an interview. The highs came when people agreed to talk. The lows came when key people said no and to not bother them again. String four or five of those no's together in a row, and it felt at times like writing a proper book about Musk was impossible.

The thing that keeps you going is that a few people do say yes and then a few more, and—interview by interview—you start to figure out how the past fits together. I'll be forever grateful to the hundreds of people who were willing to give freely of their time and especially to those who let me come back again and again with questions. There are too many of these people to list, but gracious souls—like Jeremy Hollman, Kevin Brogan, Dave Lyons, Ali Javidan, Michael Colonno, and Dolly Singh—each provided invaluable insights and abundant technical help. Heartfelt thanks go as well to Martin Eberhard and Marc Tarpenning, both of whom added crucial, rich parts to the Tesla story.

Even in this Before Elon period, Musk did permit some of his closer friends to speak with me, and they were generous with their time and intellect. That's a special thanks then to George Zachary and Shervin Pishevar, and especially to Bill Lee, Antonio Gracias, and Steve Jurvetson, who really went out of their way for Musk and for me. And I obviously owe a tremendous debt of gratitude to Justine Musk, Maye Musk, Kimbal Musk, Peter Rive, Lyndon Rive, Russ Rive, and Scott Haldeman for their time and for letting me hear some of the family stories. Talulah Riley was kind enough to let me interview her and keep prying into her husband's life. She really brought out some aspects of Musk's personality that I had not encountered elsewhere, and she helped build a much deeper understanding of him. This meant a lot to me, and, I think, it will to the readers as well.

Once Musk agreed to work with me, much of the tension that accompanied the reporting went away and was replaced by excitement. I got access to people like JB Straubel, Franz von Holzhausen, Diarmuid O'Connell, Tom Mueller, and Gwynne Shotwell, who are all among the most intelligent and compelling figures I've run into during years of reporting. I'm forever grateful for their patience explaining bits of company history and technological basics to me and for their candor. Thanks as well to Emily Shanklin, Hannah Post, Alexis Georgeson, Liz Jarvis-Shean, and John Taylor, for dealing with my constant requests and pestering, and for setting up so many interviews at Musk's companies. Mary Beth Brown, Christina Ra, and Shanna Hendriks were no longer part of Musk Land near the end of my reporting but were all amazing in helping me learn about Musk, Tesla, and SpaceX.

My biggest debt of gratitude, of course, goes to Musk. When we first started doing the interviews, I would spend the hours leading up to our chats full of nerves. I never knew how long Musk would keep participating in the project. He might have given me one interview or ten. There was real pressure to get my most crucial questions answered up front and to be to the point in my initial interviewing. As Musk stuck around, though, the conversations went longer, were more fluid, and became more enlightening. They were the things I most looked forward to every month. Whether Musk will change the course of human history in a massive way remains to be seen, but it was certainly a thrilling privilege to get to pick the brain of someone who is reaching so high. While reticent at first, once Musk committed to the project, he committed fully, and I'm thankful and honored that things turned out that way.

On a professional front, I'd like to thank my editors and coworkers over the years—China Martens, James Niccolai, John Lettice, Vindu Goel, and Suzanne Spector—each of whom taught me different lessons about the craft of writing. Special thanks go to Andrew Orlowski, Tim O'Brien, Damon Darlin, Jim Aley, and Drew Cullen, who have had the most impact on how I think about writing and reporting and are among the best mentors anyone could hope for. I must also offer up infinite thanks to Brad Wieners and Josh Tyrangiel, my bosses at *Bloomberg Businessweek*, for giving me the freedom to pursue this project. I doubt there are two people doing more to support quality journalism.

A special brand of thanks goes to Brad Stone, my colleague at the *New York Times* and then at *Businessweek*. Brad helped me shape the idea for this book, coaxed me through dark times, and was an unrivaled sounding board. I feel bad for pestering Brad so incessantly with my questions and doubts. Brad is a model colleague, always there to help anyone with advice or to step up and take on work. He's an amazing writer and an incredible friend.

Thanks as well to Keith Lee and Sheila Abichandani Sandfort. They are two of the brightest, kindest, most genuine people I know, and their feedback on the early text was invaluable.

My agent David Patterson and editor Hilary Redmon were instrumental in helping pull this project off. David always seemed to say the right thing at low moments to pick up my spirits. Frankly, I doubt the book would have happened without the encouragement and momentum he provided during the initial part of the project. Once things got going, Hilary talked me through the trickiest moments and elevated the book to an unexpected place. She tolerated my hissy fits and made dramatic improvements to the writing. It's wonderful to finish something like this and come out the other side with a pair of such good friends. Thanks so much to you both.

Last, I have to thank my family. This book turned into a living, breathing creature that made life difficult on my family for more than two years. I didn't get to see my young boys as much as I would have liked during this time, but when I did they were there with energizing smiles and hugs. I'm thankful that they both seem to have picked up an interest in rockets and cars as a result of this project. As for my wife, Melinda, well, she was a saint. From a practical perspective, this book could not have happened without her support. Melinda was my best reader and ultimate confidante. She was that best friend who knew when to try to energize me and when to let things go. Even though this book disrupted our lives for a long while, it brought us closer together in the end. I'm blessed to have such a partner, and I will forever remember what Melinda did for our family.

NOTES

1. *Journal of the Canadian Chiropractic Association*, 1995.
2. http://queensu.ca/news/alumnireview/rocket-man.
3. http://www.marieclaire.com/sex-love/relationship-issues/millionaire-starter-wife.
4. The investor Bill Lee, one of Musk's close friends, originated this phrase.
5. http://archive.wired.com/science/space/magazine/15-06/ff_space_musk?currentPage=all.
6. http://news.cnet.com/Electric-sports-car-packs-a-punch%2C-but-will-it-sell/2100-11389_3-6096377.html.
7. http://www.nytimes.com/2006/07/19/business/19electric.html.
8. A southern gentleman, Currie could never get used to Musk's swearing—"he curses like a sailor and does it in mixed company"—or the way he would churn through prized talent. "He'd search through the woods, turn over every rock and dig through brambles to find the one person with the specific expertise and skill he wanted," Currie said. "Then, that guy would be gone three months to a year later if he didn't agree with Elon." Currie, though, remembers Musk as inspirational. Even as Tesla's funds dwindled, Musk urged the employees to do their jobs well and vowed to give them what they needed to be successful. Currie, like many people, also found Musk's work ethic astonishing. "I would be in Europe or China and send him an email at two thirty in the morning his time," Currie said. "Five minutes later, I'd get an answer back. It's just unbelievable to have support on that level."

9. http://www.mercurynews.com/greenenergy/ci_7641424.

10. http://www.telegraph.co.uk/culture/3666994/One-more-giant-leap.html.

11. http://www.sia.org/wp-content/uploads/2013/06/2013_SSIR_Final.pdf.

12. Another moment like this occurred in late 2010 during a launch attempt in Florida. One of the SpaceX technicians had left a hatch open overnight at the launchpad, which allowed rain to flood a lower-level computing room. The water caused major issues with SpaceX's computing equipment, and another technician had to fly out from California right away with Musk's American Express card in hand to fix the emergency in the days leading up to the launch.

   The SpaceX engineers bought new computing gear right away and set it up in the room. They needed to run the equipment through standard tests to make sure it could maintain a certain voltage level. It was late at night on a Sunday, and they couldn't get access on short notice to a device that could simulate the high electrical load. One of the engineers improvised by going to a hardware store where he bought twenty-five headlamps for golf carts. The SpaceX crew strung them all together back at the launchpad and hung them from a wall. They then put on their sunglasses and lit everything up, knowing that if a power supply for the computing equipment could survive this test, it would be okay for the flight. The process was repeated for numerous power supplies, and the team worked from 9 P.M. that night until 7 A.M. and finished in time to keep the launch on track.

13. http://www.space.com/15874-private-dragon-capsule-space-station-arrival.html.

14. At the conclusion of the debate, Musk and I exchanged a couple of emails. He wrote, "Oil and gas is firmly in the Romney camp and they are feeding his campaign these talking points. Until recently, they didn't care about Tesla, as they thought we would fail.

   "Ironically, it is because they are starting to think Tesla might not fail that they are attacking us. The reason is that society has to function, so the less there seems to be a viable alternative to burning hydrocarbons, the less pressure there is to curb carbon emissions. If an electric car succeeds, it spoils that argument.

"Overall though, I think it is great that he mentioned us :) 'Romney Tesla' is one of the top Google searches!"

I reached out to Romney's camp months later, as sales of Tesla's soared, to see if he wanted to change his position but was rebuffed.

15. As Tesla has grown in size, the company has commanded more respect from suppliers and been able to get better parts and better deals. But outsourcing components still bothers Musk, and for understandable reasons. When it tried to ramp up production in 2013, Tesla ran into periodic issues because of its suppliers. One of them made what should have been an inconsequential 12-volt lead acid battery that handled a few auxiliary functions in the car. Tesla bought the part from an American supplier, which in turn outsourced the part from a company in China, which in turn outsourced the part from a company in Vietnam. By the time the battery arrived at Tesla's factories, it didn't work, adding cost and delays during a crucial period in the Model S's history. It's situations like these that typically result in Tesla playing a much more active role with its suppliers when compared to other automakers. For something like an ABS braking controller, Tesla will work hand-in-hand with its supplier—in this case Bosch—to tune the hardware and software for the Model S's specific characteristics. "Most companies just hand their cars over to Bosch, but Tesla goes in with a software engineer," said Ali Javidan. "We had to change their mind-set and let them know we wanted to work on a very deep level."

16. Tesla does seem to promote an obsession with safety that's unmatched in the industry. J. B. Straubel explained the company's thinking as follows: "With the safety stuff, it seems like car companies have evolved to a place where their design objectives are set by whatever is regulated or has been standardized. The rule says, 'Do this and nothing more.' That is amazingly boring engineering. It leaves you maybe fiddling with the car's shape or trying to make it a bit faster. We have more crumple

zones, better deceleration, a lower center of gravity. We went in wondering, 'Can we make this car twice as safe as anything else on the road?'"

17. Othmer has lined up to be the lucky owner of the first Roadster II.

    Musk has developed an unconventional policy to determine the order in which cars are sold. When a new car is announced and its price is set, a race begins in which the first person to hand Musk a check gets the first car. With the Model S, Steve Jurvetson, a Tesla board member, had a check at the ready in his wallet and slid it across the table to Musk after spying details on the Model S in a packet of board meeting notes.

    Othmer caught a *Wired* story about a planned second version of the Roadster and emailed Musk right away. "He said, 'Okay, I will sell it to you, but you have to pay two hundred thousand dollars right now.'" Othmer agreed, and Tesla had him come to the company's headquarters on a Sunday to sign some paperwork, acknowledging the price of the car and the fact that the company didn't quite know when it would arrive or what its specifications would be. "My guess is that it will be the fastest car on the road," Othmer said. "It'll be four-wheel drive. It's going to be insane. And I don't really think that will be the real price. I just don't think Elon wanted me to buy it."

18. Musk wondered if Better Place came up with battery swapping as a plan after its CEO, Shai Agassi, heard about the technology during a tour of the Tesla factory

19. Musk had made a number of art cars over the years at Burning Man, including an electric one shaped like a rocket. In 2011, he also received a lot of grief from the *Wall Street Journal* for having a high-end camp. "Elon Musk, chief executive of electric-car maker Tesla Motors and co-founder of eBay Inc.'s PayPal unit, is among those eschewing the tent life," the paper wrote. "He is paying for an elaborate compound consisting of eight recreational vehicles and trailers stocked with food, linens, groceries and other essentials for himself and his friends and family, say employees of the outfitter, Classic Adventures RV. . . . Classic is one of the festival's few approved vendors. It charges $5,500 to $10,000 per RV for its Camp Classic Concierge packages like

Mr. Musk's. At Mr. Musk's RV enclave, the help empties septic tanks, brings water and makes sure the vehicles' electricity, refrigeration, air conditioning, televisions, DVD players and other systems are ship shape. The staff also stocked the campers with Diet Coke, Gatorade and Cruzan rum." Once the story hit, Musk's group tried to move to a new, undisclosed location.

20. http://www.sandia.gov/~jytsao/Solar%20FAQs.pdf.

21. Tesla employees have been known to sneak across the street to the campus of the software maker SAP and to take advantage of its sumptuous, subsidized cafes.

22. Shotwell talks about going to Mars as much as Musk and has dedicated her life to space exploration. Straubel has demonstrated the same type of commitment with electric vehicles and can sound a lot like Musk at times. "We are not trying to corner the market on EVs," Straubel said. "There are 100 million cars built per year and 2 billion already out there. Even if we got to 5 or 10 percent of the market, that does not solve the world's problems. I am bullish we will keep up with demand and drive the whole industry forward. Elon is committed to this."

23. Page presented one of his far-out ideas to me as follows: "I was thinking it would be pretty cool to have a prize to fund a project where someone would have to send something lightweight to the moon that could sort of replicate itself. I went over to the NASA operation center here at AMES in Mountain View when they were doing a mission and literally flying a satellite into the south pole of the moon. And they like hurled this thing into the moon at a high velocity and then it exploded and it sent matter out into space. And then they looked at that with telescopes, and they discovered water on the south pole of the moon, which sounds really exciting. I started thinking that if there's a lot of water on the south pole of the moon, you can make rocket fuel from the hydrogen and oxygen. The other cool thing about the south pole is like it almost always gets sun. There's like places high up that get sun and there's places that are kind of in the craters that are very cold. So you have like a lot of energy then where you could run solar cells. You could almost run like a steam turbine there. You have rocket fuel ingredients, and you

have solar cells that can be powered by sun, and you could probably run a power plant turbine. Power plant turbines aren't that heavy. You could send that to the moon. You have like a gigawatt of power on the moon and make a lot of rocket fuel. It would make a good prize project. You send something to the moon that weights five pounds and have it make rocket fuel so that you could launch stuff off the moon or have it make a copy of itself, so that you can make more of them."